Student Workbook
to Accompany

Drafting & Design for Architecture & Construction

Ninth Edition

Dana J. Hepler
Principal, Hepler Associates PC
New York City and Massapequa Park NY
Adjunct Professor of Architecture
New York Institute of Technology, Old Westbury, NY

Paul Ross Wallach
Architecture Instructor, Canada College
Technical Writer and Consultant,
Burlingham, CA

Donald E. Hepler
President, Technical Writing and Design Service Inc.
Somers, CT

Australia • Brazil • Japan • Korea • Mexico • Singapore • Spain • United Kingdom • United States

Drafting & Design for Architecture & Construction Workbook, Ninth Edition
Dana J. Hepler, Paul R. Wallach, and Donald E. Hepler

Vice President, Editorial: Dave Garza

Director of Learning Solutions: Sandy Clark

Associate Acquisitions Editor: Kathryn Hall

Managing Editor: Larry Main

Senior Product Manager: Sharon Chambliss

Editorial Assistant: Cristopher Savino

Vice President, Marketing: Jennifer Baker

Marketing Director: Deborah Yarnell

Associate Marketing Manager: Jillian Borden

Production Director: Wendy Troeger

Production Manager: Mark Bernard

Content Project Manager: Sara Dovre Wudali

Art Director: Casey Kirchmayer

Technology Project Manager: Christopher Catalina

© 2013 Delmar, Cengage Learning

ALL RIGHTS RESERVED. No part of this work covered by the copyright herein may be reproduced, transmitted, stored, or used in any form or by any means graphic, electronic, or mechanical, including but not limited to photocopying, recording, scanning, digitizing, taping, Web distribution, information networks, or information storage and retrieval systems, except as permitted under Section 107 or 108 of the 1976 United States Copyright Act, without the prior written permission of the publisher.

For product information and technology assistance, contact us at
Cengage Learning Customer & Sales Support, 1-800-354-9706
For permission to use material from this text or product,
submit all requests online at **www.cengage.com/permissions**.
Further permissions questions can be e-mailed to
permissionrequest@cengage.com

Library of Congress Control Number: 2011939833

ISBN-13: 978-1-1111-2815-9

ISBN-10: 1-1111-2815-4

Delmar
5 Maxwell Drive
Clifton Park, NY 12065-2919
USA

Cengage Learning is a leading provider of customized learning solutions with office locations around the globe, including Singapore, the United Kingdom, Australia, Mexico, Brazil, and Japan. Locate your local office at: **international.cengage.com/region**

Cengage Learning products are represented in Canada by Nelson Education, Ltd.

To learn more about Delmar, visit **www.cengage.com/delmar**

Purchase any of our products at your local college store or at our preferred online store **www.cengagebrain.com**

Notice to the Reader
Publisher does not warrant or guarantee any of the products described herein or perform any independent analysis in connection with any of the product information contained herein. Publisher does not assume, and expressly disclaims, any obligation to obtain and include information other than that provided to it by the manufacturer. The reader is expressly warned to consider and adopt all safety precautions that might be indicated by the activities described herein and to avoid all potential hazards. By following the instructions contained herein, the reader willingly assumes all risks in connection with such instructions. The publisher makes no representations or warranties of any kind, including but not limited to, the warranties of fitness for particular purpose or merchantability, nor are any such representations implied with respect to the material set forth herein, and the publisher takes no responsibility with respect to such material. The publisher shall not be liable for any special, consequential, or exemplary damages resulting, in whole or part, from the readers' use of, or reliance upon, this material.

Printed in the United States of America
1 2 3 4 5 6 7 15 14 13 12

CONTENTS

Introduction iv

Section I—Drawing Exercises.................................... 1

Exercise 1	Architectural History and Styles	3
Exercise 2	Fundamentals of Design	5
Exercise 3	Scales and Measurements	7
Exercise 4	Drafting Conventions and Procedures	13
Exercise 5	Introduction to Computer-Aided Drafting and Design	26
Exercise 6	Environmental Design Factors	30
Exercise 7	Indoor Living Areas	34
Exercise 8	Outdoor Living Areas	36
Exercise 9	Traffic Areas and Patterns	39
Exercise 10	Kitchens	41
Exercise 11	Service Areas	47
Exercise 12	Sleeping Areas	49
Exercise 13	Site Development Plans	53
Exercise 14	Designing Floor Plans	59
Exercise 15	Drawing Floor Plans	66
Exercise 16	Designing Elevations	71
Exercise 17	Drawing Elevations	78
Exercise 18	Sectional Detail and Cabinetry Drawings	93
Exercise 19	Pictorial Drawings	95
Exercise 20	Architectural Renderings	105
Exercise 21	Architectural Models	108
Exercise 22	Principles of Construction	114
Exercise 23	Foundations and Fireplace Structures	117
Exercise 24	Wood-Frame Systems	126

Exercise 25	Masonry and Concrete Systems	128
Exercise 26	Steel and Reinforced-Concrete Systems	130
Exercise 27	Disaster Prevention Design	131
Exercise 28	Floor Framing Drawings	133
Exercise 29	Wall Framing Drawings	140
Exercise 30	Roof Framing Drawings	142
Exercise 31	Electrical Design and Drawings	149
Exercise 32	Comfort-Control Systems (HVAC)	154
Exercise 33	Plumbing Drawings	158
Exercise 34	Drawing Coordination and Checking	161
Exercise 35	Schedules and Specifications	169
Exercise 36	Building Costs and Financial Planning	172
Exercise 37	Codes and Legal Documents	174

Section II—Drawing Templates 179

T-1: Floor Plan Rooms 180
T-2: Living Area 181
T-3: Bedroom And Service Area 183
T-4: Living And Sleeping Areas—Elevations 184
T-5: Service Area—Elevations 185

Section III—Chapter Review Tests 187

INTRODUCTION

The student workbook accompanies the ninth edition of *Drafting & Design for Architecture & Construction* and contains exercises, problems, and tests designed to provide experiences that reinforce and strengthen student knowledge of the principles and practices of architecture and construction drafting and design.

The workbook contains three sections: Section I, Drawing Exercises; Section II, Drawing Templates; Section III, Chapter Review Tests.

SECTION 1
Drawing Exercises

The exercises contained in this section provide students with preliminary practice in problem solving through drawing and/or sketching. Completing these exercises prior to the preparation of an actual architectural design helps clarify concepts and procedures and eliminates time-consuming errors during the design process.

The exercises follow the chapter sequence of the text and may also be used as achievement tests, out-of-class assignments, classroom activities, or pretests. The problems range from the very simple, which can be quickly completed, to those that require more prolonged study and skill. All exercises can be completed though sketching, board drafting, and/or computer-aided drafting. Regardless of the method used, accurate measurements, lettering, and stand sketching and/or drawing techniques should be used.

These drawing exercises are designed to augment and reinforce the exercises found at the end of each chapter in the text. Additional exercises, tests, and problems are included in the *Instructor's Guide to Drafting and Design for Architecture and Construction*. Information relating to the solutions for all problems is contained in each related chapter in the text.

EXERCISE 1-1

In the spaces provided, sketch these examples of early Greek construction details.

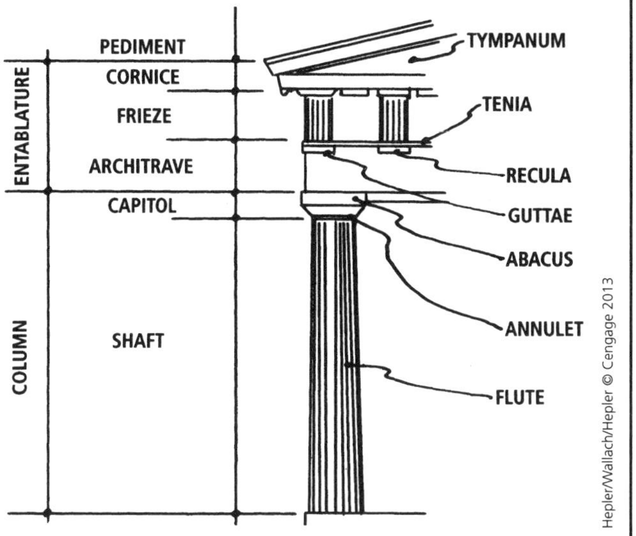

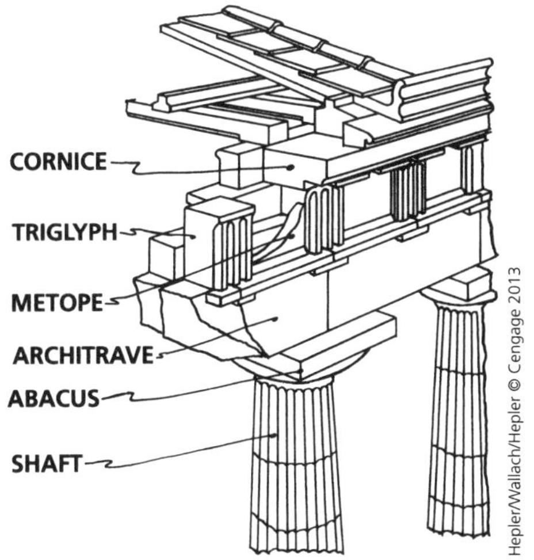

ENTABLATURE
- PEDIMENT
- CORNICE
- FRIEZE
- ARCHITRAVE

COLUMN
- CAPITOL
- SHAFT

- TYMPANUM
- TENIA
- RECULA
- GUTTAE
- ABACUS
- ANNULET
- FLUTE

- CORNICE
- TRIGLYPH
- METOPE
- ARCHITRAVE
- ABACUS
- SHAFT

PARTHENON

TITLE	DRAFTER	SCALE	DATE	DWG NO.

EXERCISE 1 Architectural History and Styles ■ 3

EXERCISE 1-2

In the spaces provided, sketch these examples of early Roman arches.

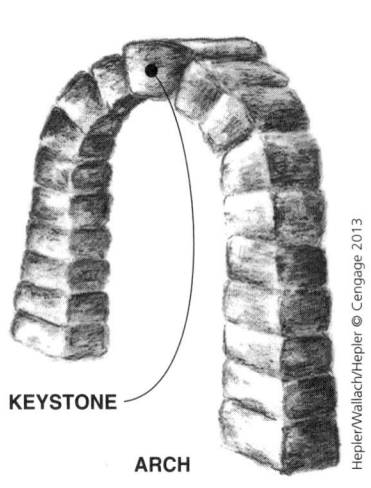

KEYSTONE
ARCH

BARREL VAULT

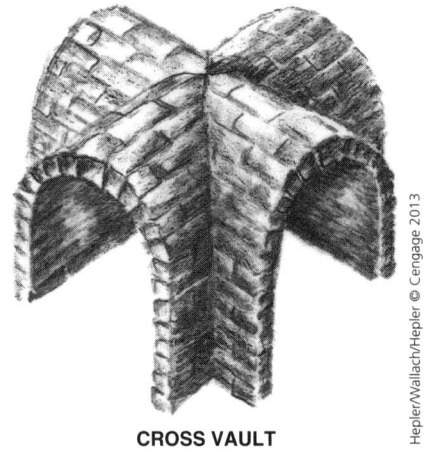

CROSS VAULT

TITLE	DRAFTER	SCALE	DATE	DWG NO.

EXERCISE 1-1

In the spaces provided, sketch these examples of early Greek construction details.

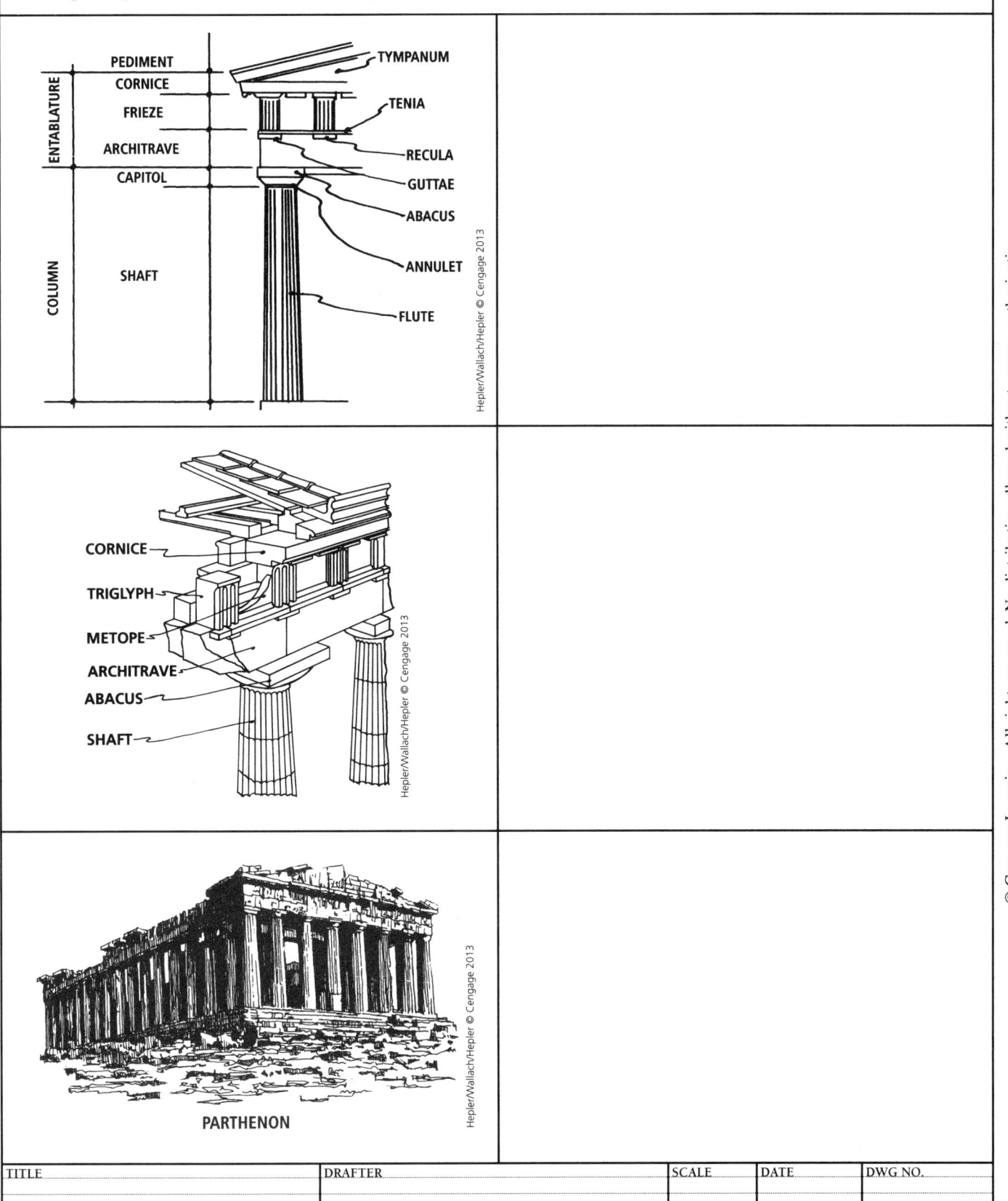

| TITLE | DRAFTER | SCALE | DATE | DWG NO. |

EXERCISE 1-2

In the spaces provided, sketch these examples of early Roman arches.

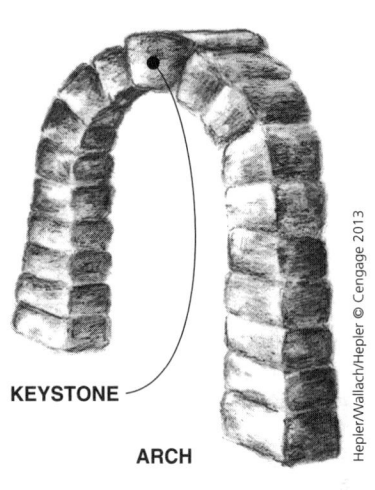
KEYSTONE
ARCH

BARREL VAULT

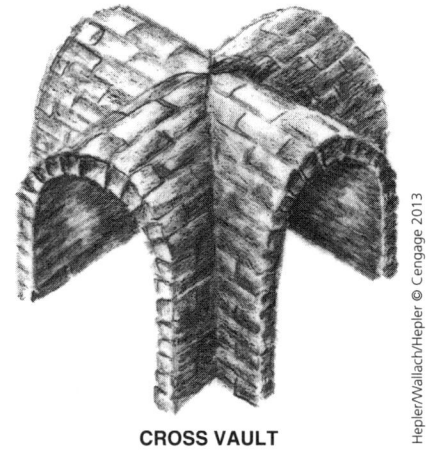
CROSS VAULT

TITLE	DRAFTER	SCALE	DATE	DWG NO.

EXERCISE 2-1

Which type of balance, symmetrical or asymmetrical, was used for this house design?
_____
In the space below, sketch a new design for this house, using as many of the same features as you can.

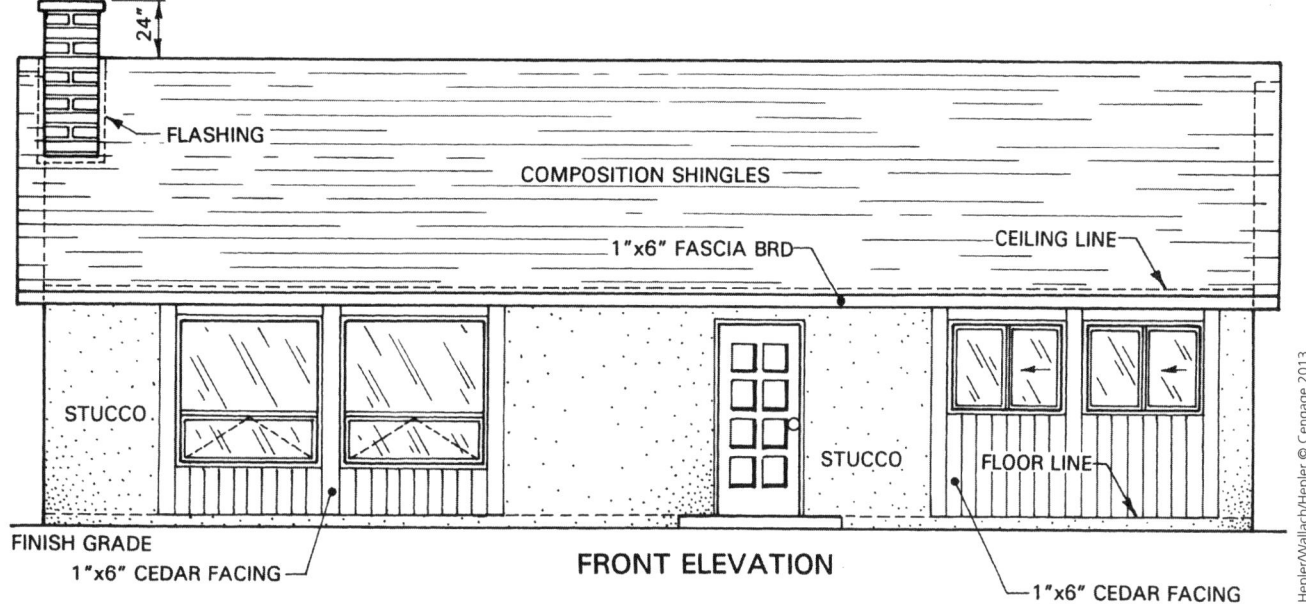

TITLE	DRAFTER	SCALE	DATE	DWG NO.

EXERCISE 2-2

Make a rough sketch of a house of your own design in the space below. Feature one or two of the elements and principles of design. When you are finished, label the elements and principles you have used.

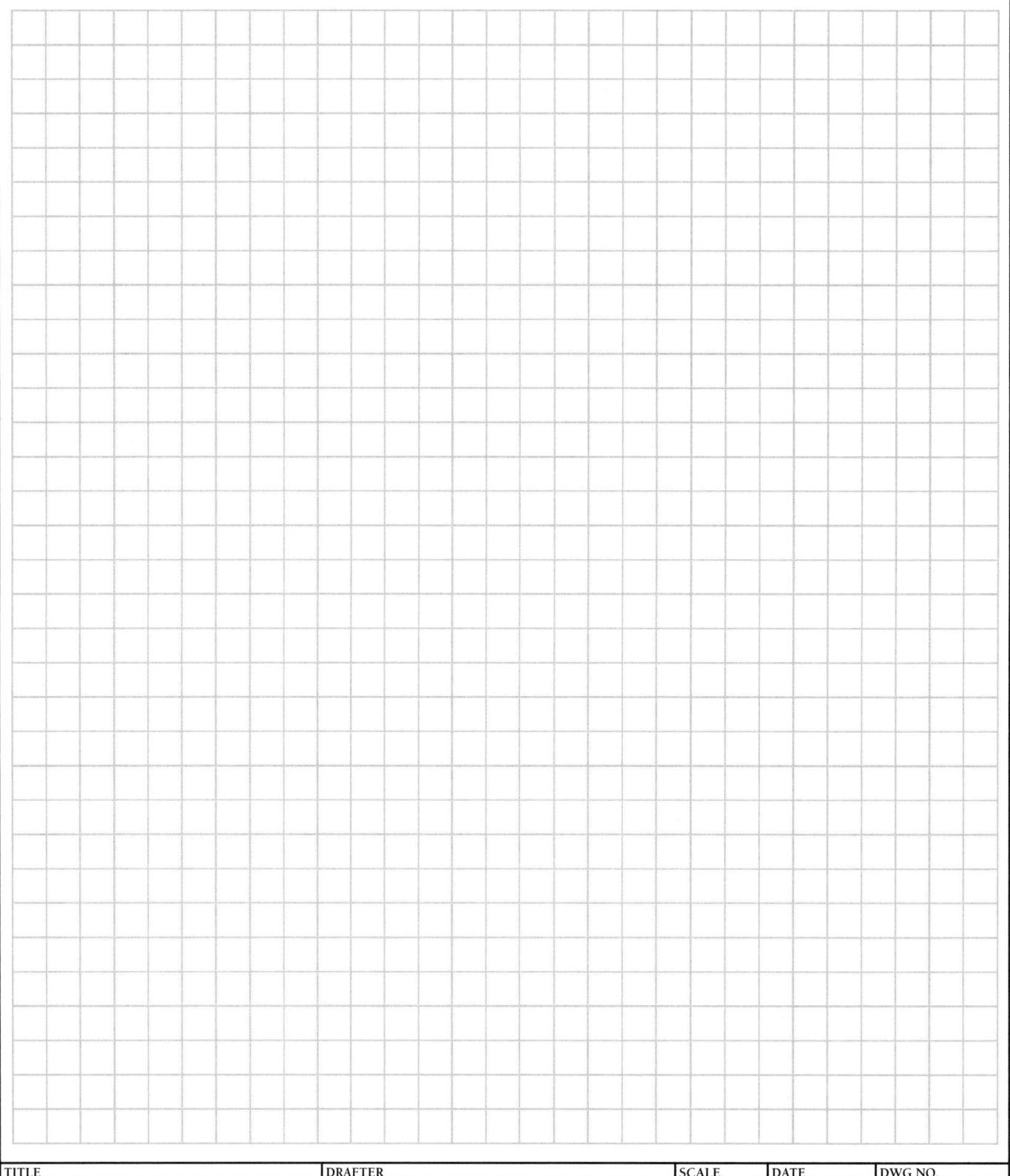

EXERCISE 3-1

Read the dimensions for the mechanical engineer's scale to the closest 1/16 inch on the scale and write them in the spaces provided. Read the dimensions for the architect's scale to the closest inch on the scale and write them in the spaces provided. Scales are not actual size.

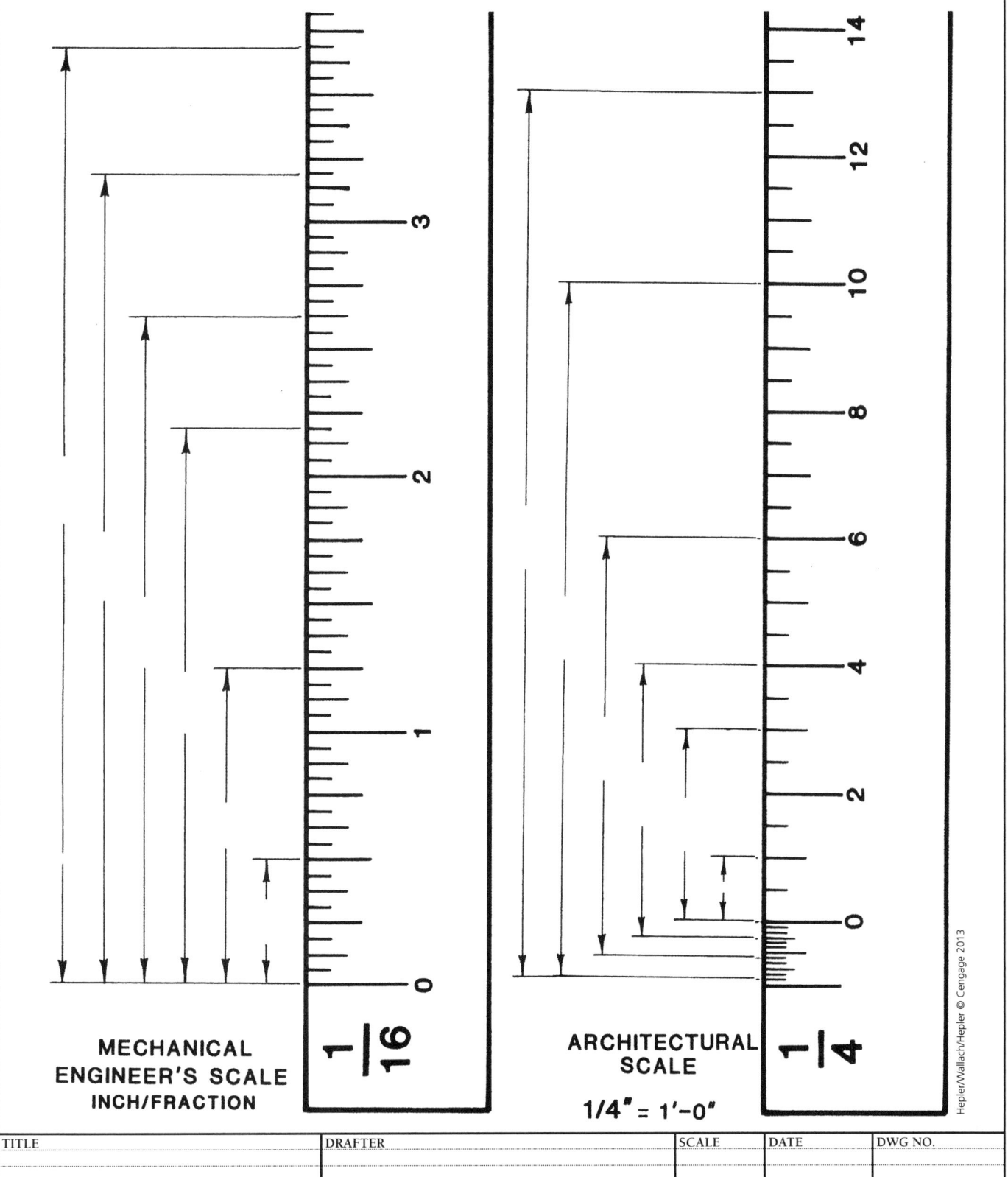

EXERCISE 3-2

Different scales are listed below. Following each scale is a distance in feet and inches. Using the scale indicated, measure and mark each distance on the line to the right.

1. 1/4" = 1'–0"   20'–0"   _____
2. 1/4" = 1'–0"   21'–0"   _____
3. 1/4" = 1'–0"   11'–6"   _____
4. 1/4" = 1'–0"   15'–6"   _____
5. 1/4" = 1'–0"   8'–9"   _____
6. 1/4" = 1'–0"   7'–3"   _____
7. 1/4" = 1'–0"   10'–1"   _____
8. 1/8" = 1'–0"   40'–11"   _____
9. 1/8" = 1'–0"   25'–0"   _____
10. 1/8" = 1'–0"   12'–6"   _____
11. 1" = 1'–0"   3'–5"   _____
12. 1" = 1'–0"   4'–9 1/2" _____
13. 1/2" = 1'–0"   9'–6"   _____
14. 1/2" = 1'–0"   5'–2"   _____
15. 1/2" = 1'–0"   7'–11"   _____
16. 1/2" = 1'–0"   6'–5"   _____
17. 3/4" = 1'–0"   5'–10 1/2" _____
18. 3/8" = 1'–0"   12'–11"   _____
19. 3/8" = 1'–0"   6'–2"   _____
20. 3/16" = 1'–0" 27'–10"   _____
21. 3/16" = 1'–0" 15'–2"   _____
22. 3/32" = 1'–0" 50'–0"   _____
23. 3/32" = 1'–0" 42'–8"   _____
24. 1 1/2" = 1'–0" 2'–3 1/2" _____
25. 1 1/2" = 1'–0" 1'–9 1/4" _____

TITLE	DRAFTER	SCALE	DATE	DWG NO.

EXERCISE 3 Scales and Measurements ■ **9**

EXERCISE 3-3

Multiple-choice questions: Select the correct answers and write the letter in the space at the right.

1. Standard scales use
   a. miles.
   b. inches and their divisions.
   c. centimeters and millimeters.
   d. kilometers.

   1. _____

2. Metric scales use
   a. miles.
   b. inches and their divisions.
   c. centimeters and millimeters.
   d. kilometers.

   2. _____

3. Construction details use
   a. plot plans.
   b. small scale.
   c. large scale.
   d. metric scales only.

   3. _____

4. Plot plans use
   a. small scale.
   b. larger dimensions.
   b. letters.
   d. variables.

   4. _____

Measure the length of each line to the closest inch on the scale. Use a scale of 1/8″ = 1–0″.

1. _____ 1. _____
2. _____   2. _____
3. _____   3. _____
4. _____   4. _____
5. _____   5. _____
6. _____   6. _____
7. _____   7. _____
8. _____ 8. _____
9. _____   9. _____
10. _____   10. _____

Scale: 1/4″ = 1′–0″

1. _____ 1. _____
2. _____   2. _____
3. _____ 3. _____
4. _____   4. _____
5. _____   5. _____
6. _____   6. _____
7. _____ 7. _____
8. _____   8. _____
9. _____   9. _____
10. _____   10. _____

TITLE	DRAFTER	SCALE	DATE	DWG NO.

EXERCISE 3-4

Measure the length of each line to the nearest inch on the scale.

Scale: 1/2" = 1'–0"

1. _____   1. _____
2. _____   2. _____
3. _____   3. _____
4. _____   4. _____
5. _____   5. _____
6. _____   6. _____
7. _____   7. _____
8. _____   8. _____
9. _____   9. _____
10. _____   10. _____
11. _____   11. _____
12. _____   12. _____
13. _____   13. _____
14. _____   14. _____
15. _____   15. _____

Measure to closest quarter inch on the scale.

Scale: 1/4" = 1'–0"

1. _____   1. _____
2. _____   2. _____
3. _____   3. _____
4. _____   4. _____
5. _____   5. _____
6. _____   6. _____
7. _____   7. _____
8. _____   8. _____
9. _____   9. _____
10. ____   10. _____
11. _____   11. _____
12. _____   12. _____
13. _____   13. _____
14. _____   14. _____
15. _____   15. _____
16. _____   16. _____
17. _____   17. _____
18. _____   18. _____
19. _____   19. _____

TITLE	DRAFTER	SCALE	DATE	DWG NO.

EXERCISE 4-1

Using the space provided, practice the line conventions shown here and on the next page. If you need more space, use a separate sheet of paper.

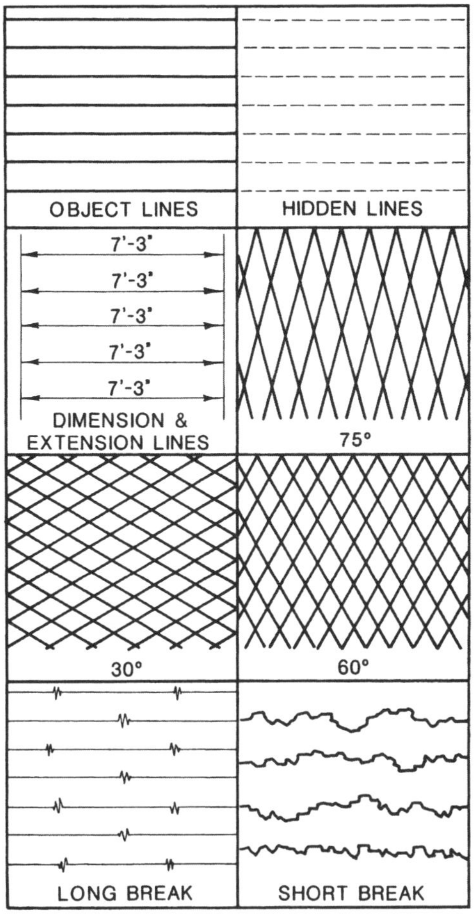
OBJECT LINES
HIDDEN LINES
7'-3"
7'-3"
7'-3"
7'-3"
7'-3"
DIMENSION & EXTENSION LINES
75°
30°
60°
LONG BREAK
SHORT BREAK

Hepler/Wallach/Hepler © Cengage 2013

TITLE	DRAFTER	SCALE	DATE	DWG NO.

EXERCISE 4-1 (CONTINUED)

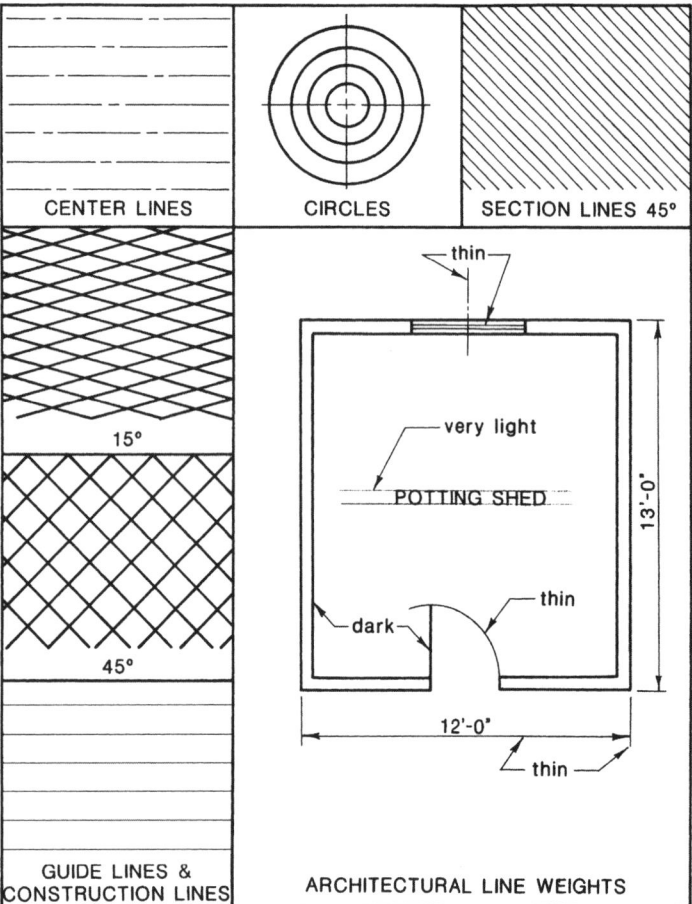

| TITLE | DRAFTER | SCALE | DATE | DWG NO. |

EXERCISE 4-2, PART A

Using drafting instruments, copy the geometric forms in the spaces provided.

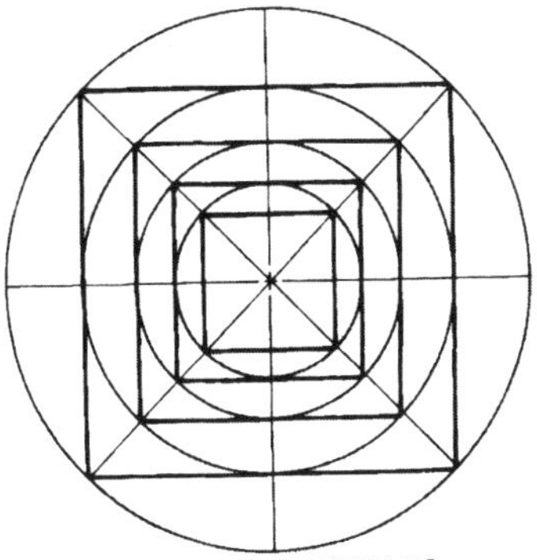

SQUARES & CIRCLES

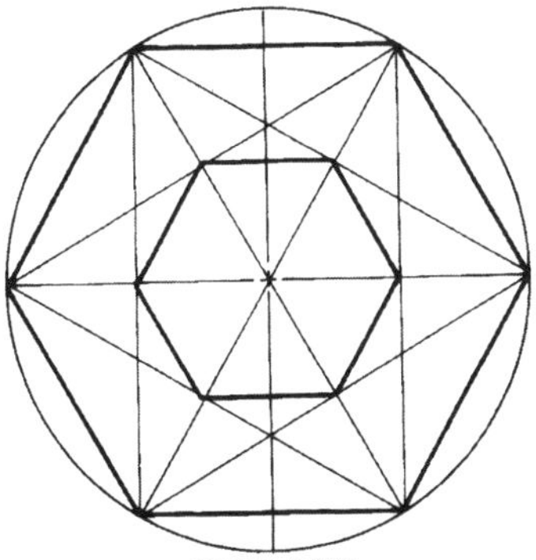

HEXAGON

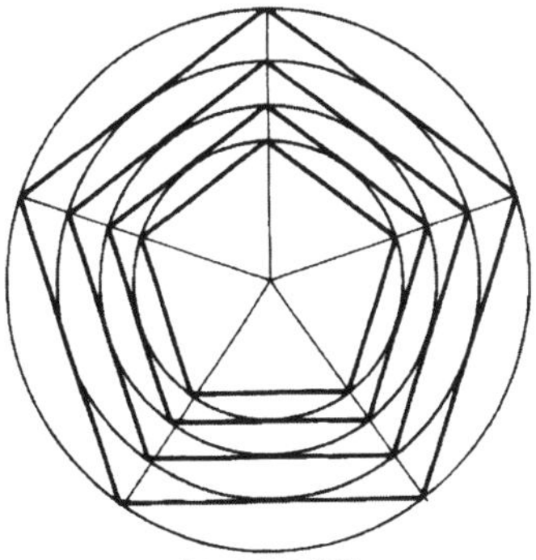

PENTAGON

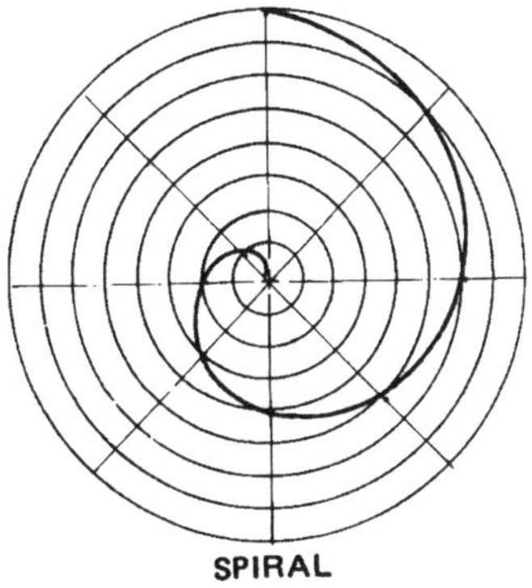

SPIRAL

| TITLE | DRAFTER | SCALE | DATE | DWG NO. |

EXERCISE 4-2, PART B

Using drafting instruments, practice making geometric designs in the space below.

TITLE	DRAFTER	SCALE	DATE	DWG NO.

EXERCISE 4-5

Practice instrument line work. Try to improve with each attempt.

Vertical and horizontal	1	2	3
30°	1	2	3
45°	1	2	3
60°	1	2	3
Circles	1	2	3

TITLE	DRAFTER	SCALE	DATE	DWG NO.

EXERCISE 4-6

Practice drawing the different line standards.

1. Border lines (very heavy)

2. Cutting plane lines (very heavy)

3. Visible lines (heavy)

4. Hidden lines (thin)

5. Center lines (thin)

6. Section lines (thin)

7. Leaders with arrowheads (thin)

8. Extension and dimension lines (thin)

9. Long break lines (thin)

10. Short break lines (heavy)

TITLE	DRAFTER	SCALE	DATE	DWG NO.

EXERCISE 4-7

Complete the title blocks.

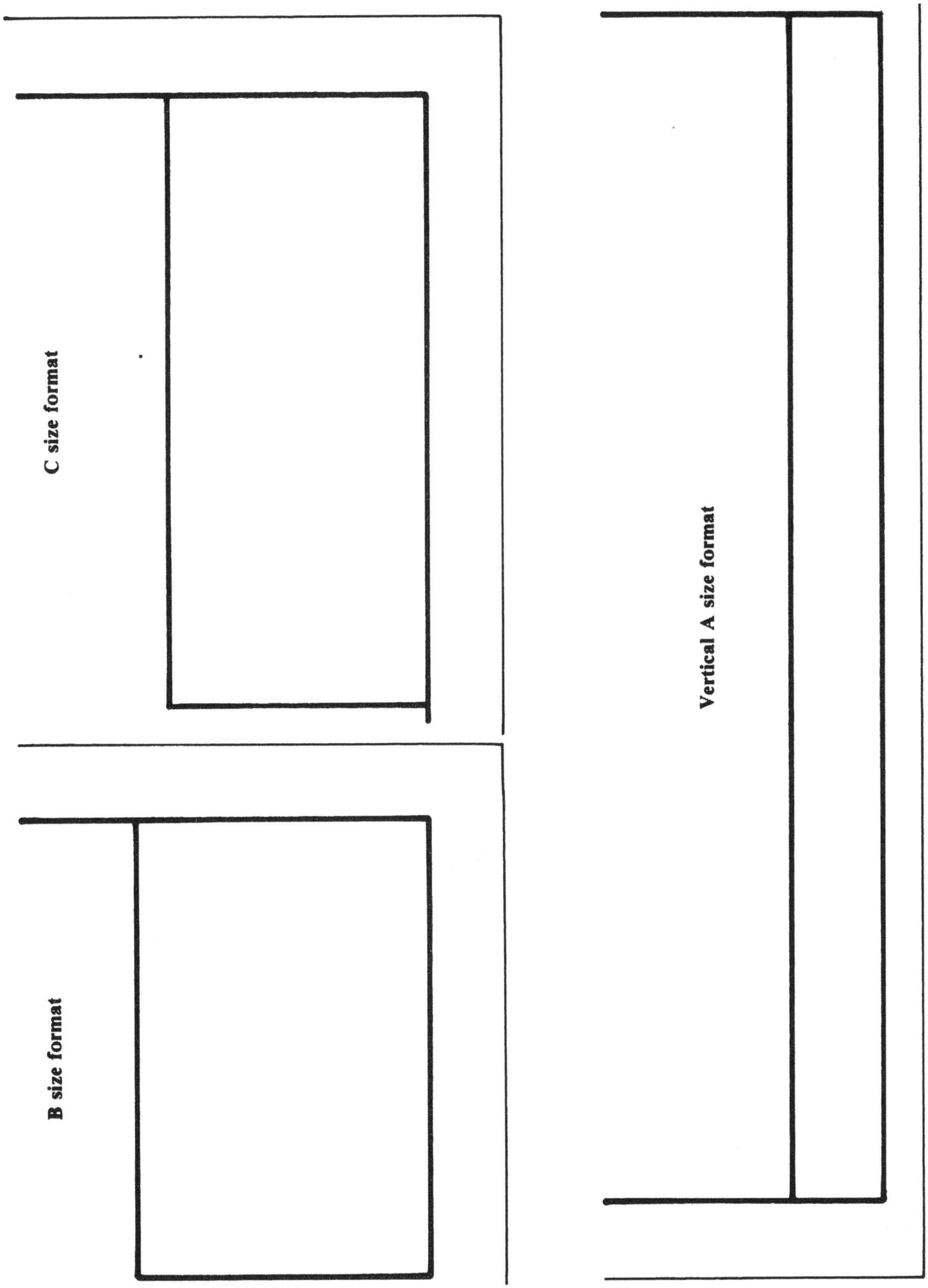

C size format

Vertical A size format

B size format

TITLE	DRAFTER	SCALE	DATE	DWG NO.

EXERCISE 4-8

On the lines below and on the next page, practice making these different styles of lettering.

ABCDEFGHIJKLMNOPQRSTUVWXYZ & 12345
ABCDEFGHIJKLMNOPQRSTUVWXYZ & 1234567890
ABCDEFGHIJKLMNOPQRSTUVWXYZ123456789

Hepler/Wallach/Hepler © Cengage 2013

EXERCISE 4-8 (CONTINUED)

| TITLE | DRAFTER | SCALE | DATE | DWG NO. |

EXERCISE 4-9

Practice duplicating these alphabet styles on the lines below.

SLANT (68 DEGREES) SINGLE STROKE GOTHIC
ABCDEFGHIJKLMNOPQRSTUVWXYZ 1234567890

ARCHITECTURAL STYLE
ABCDEFGHIJKLMNOPQRSTUVWXYZ 1234567890

TITLE	DRAFTER	SCALE	DATE	DWG NO.

EXERCISE 4-10

Practice duplicating the alphabet shown on the lines below.

SINGLE STROKE GOTHIC LETTERS AND NUMBERS
ABCDEFGHIJKLMNOPQRSTUVWXYZ   0123456789
ABCDEFGHIJKLMNOPQRSTUVWXYZ   *0123456789*

Hepler/Wallach/Hepler © Cengage 2013

| TITLE | DRAFTER | SCALE | DATE | DWG NO. |

EXERCISE 5-1

Complete the data table for the plot plan by identifying the coordinate points. The points for line 1 have been done for you.

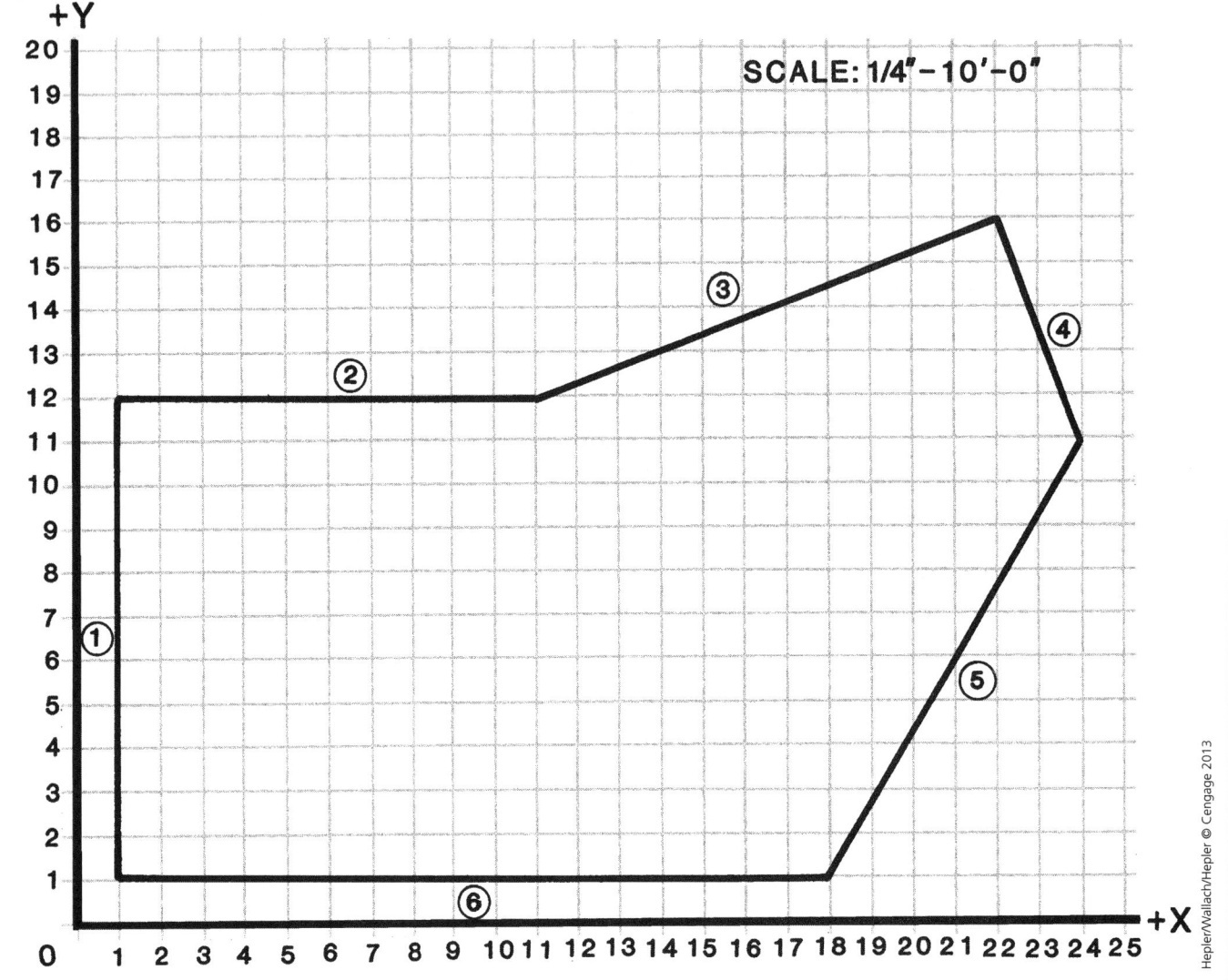

EXERCISE 5-2

Using the coordinates on the data table, draw the lines on the grid below. Zero is indicated for you.

DATA TABLE					
LINE	X	Y	go to	X	Y
1	3	2	→	3	7
2	2	6	→	8	13
3	8	13	→	13	6
4	12	7	→	12	2
5	12	2	→	3	2
6	6	2	→	6	6
7	6	6	→	9	6
8	9	6	→	9	2

TITLE	DRAFTER	SCALE	DATE	DWG NO.

EXERCISE 5-3

Complete the data tables by referring to the drawings.

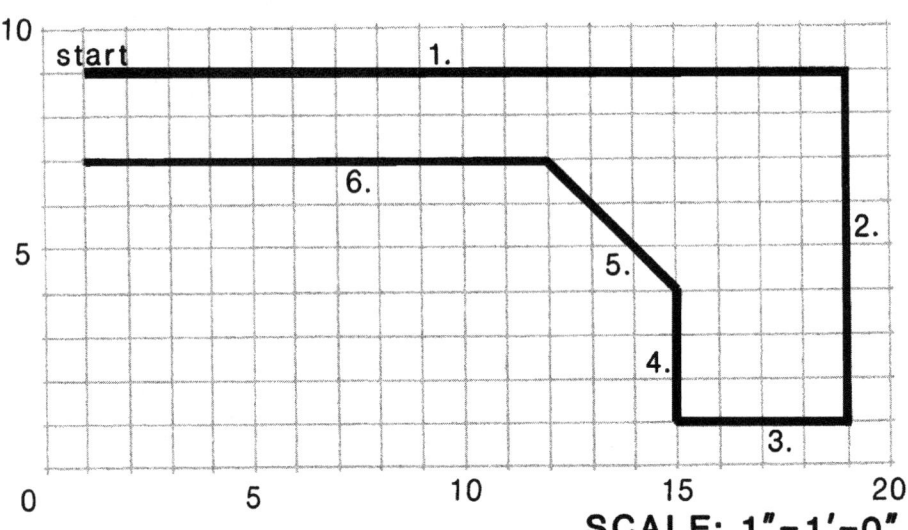

line	X	Y	go to	X	Y
1	1	9		19	9
2	19	9			
3					
4					
5					
6					

SCALE: 1"=1'-0"

line	X	Y	go to	X	Y
1	1	18		19	18
2	19	18			
3					
4					
5					
6					
7					
8					
new start					
9					
new start					
10					

SCALE: 1"=1'-0

TITLE	DRAFTER	SCALE	DATE	DWG NO.

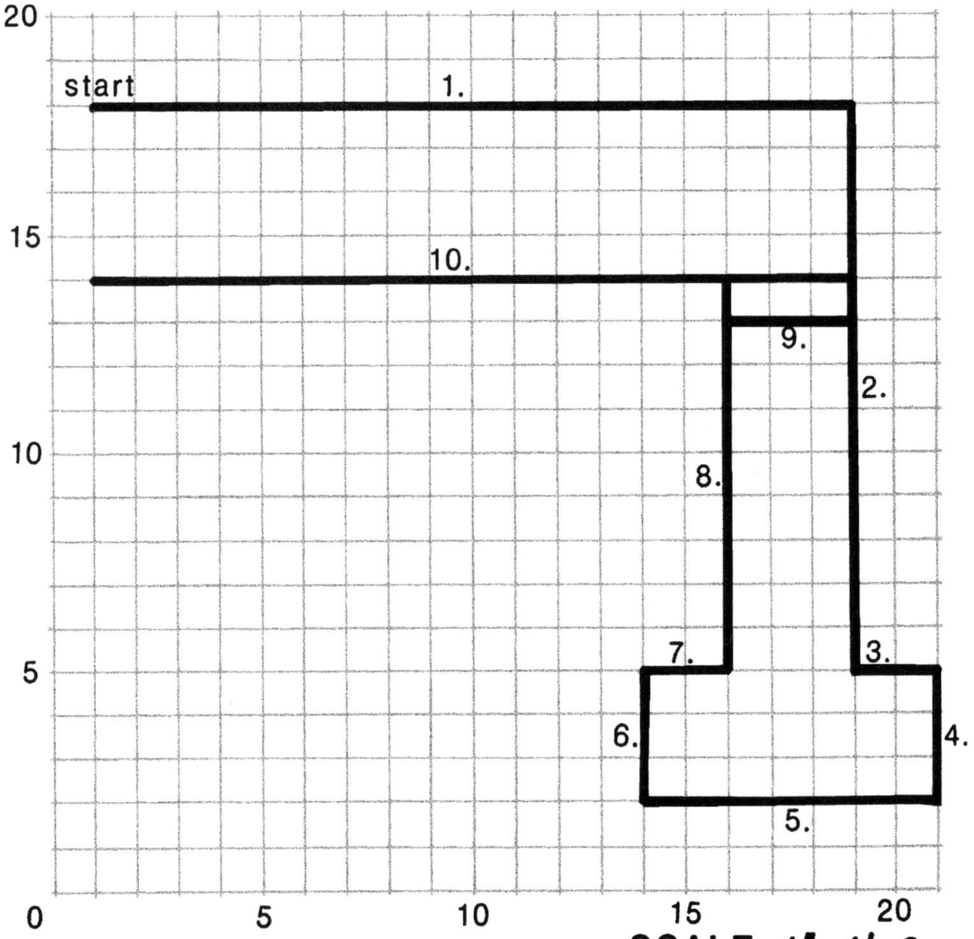

EXERCISE 5-4
Complete the drawings by referring to the data tables.

line	X	Y	go to	X	Y
1	0	13		20	13
2	20	13		20	8
3	20	8		0	8
4	0	8		0	13
	new start				
5	0	8.5		20	8.5
	new start				
6	1	8		1	1
7	1	1		19	1
8	19	1		19	8
	new start				
9	3	4		7	4
10	7	4		7	7
11	7	7		3	7
12	3	7		3	4
	new start				
13	13	4		17	4
14	17	4		17	7
15	17	7		13	7
16	13	7		13	4

SCALE: 1/4"-1'-0"

line	X	Y	go to	X	Y
1	2	8		2	0
2	2	0		18	0
3	18	0		18	8
	new start				
4	0	7		10	12
5	10	12		20	7
	new start				
6	5	0		5	7
7	5	7		8	7
8	8	7		8	0

SCALE: 1/4"-1'-0"

TITLE	DRAFTER	SCALE	DATE	DWG NO.

EXERCISE 6-1

Suppose this piece of property were in your community. How would you develop it to preserve its natural beauty? It measures approximately 180′ wide by 100′ deep. (The scale used here is approximately 1″ = 10′.) List or sketch your ideas in the space provided.

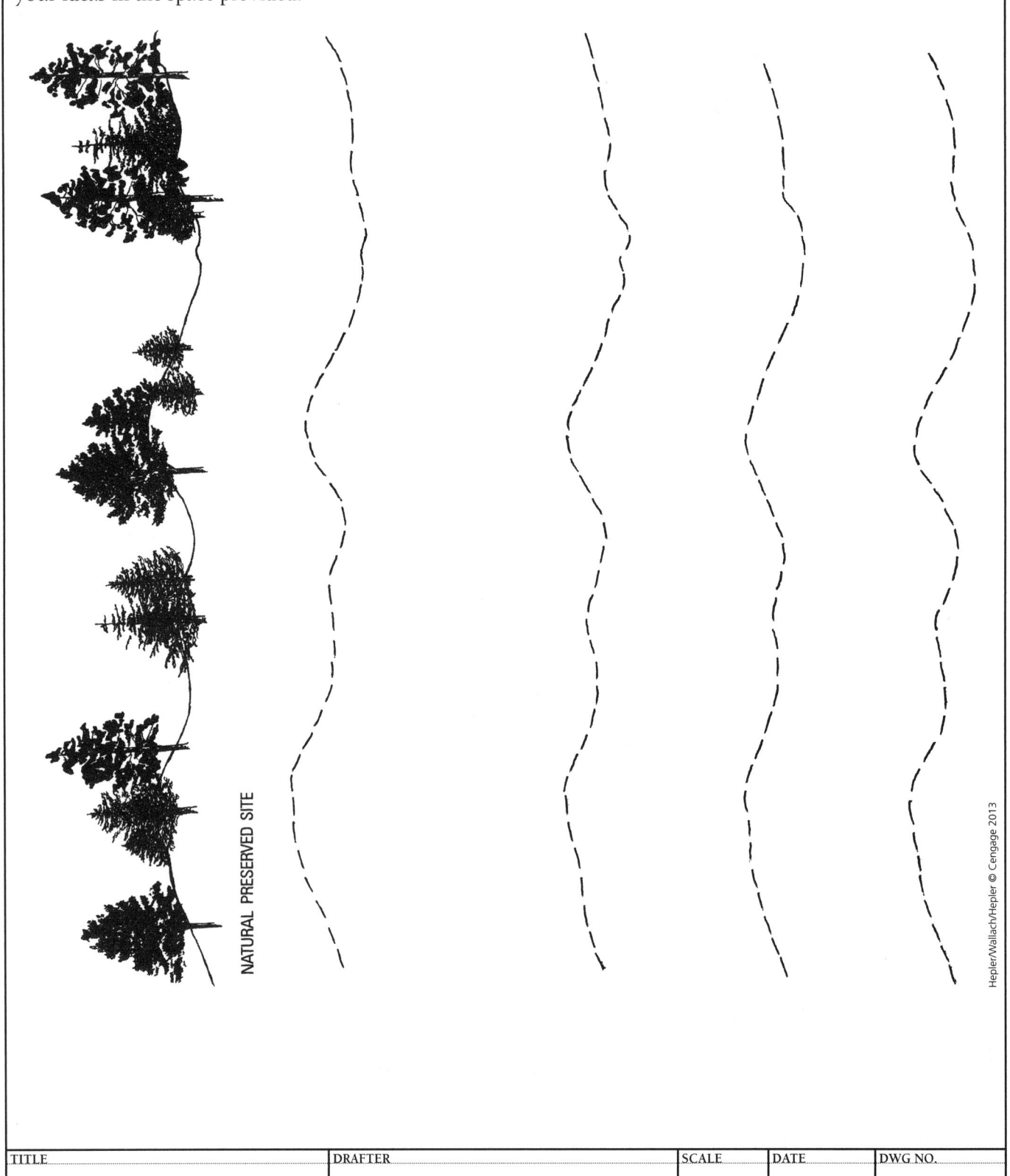

| TITLE | DRAFTER | SCALE | DATE | DWG NO. |

EXERCISE 6-2

Use the space below to design a building site that you would like to develop as a residential site.

| TITLE | DRAFTER | SCALE | DATE | DWG NO. |

EXERCISE 6-3

Arrange the bubbles for an efficient orientation.

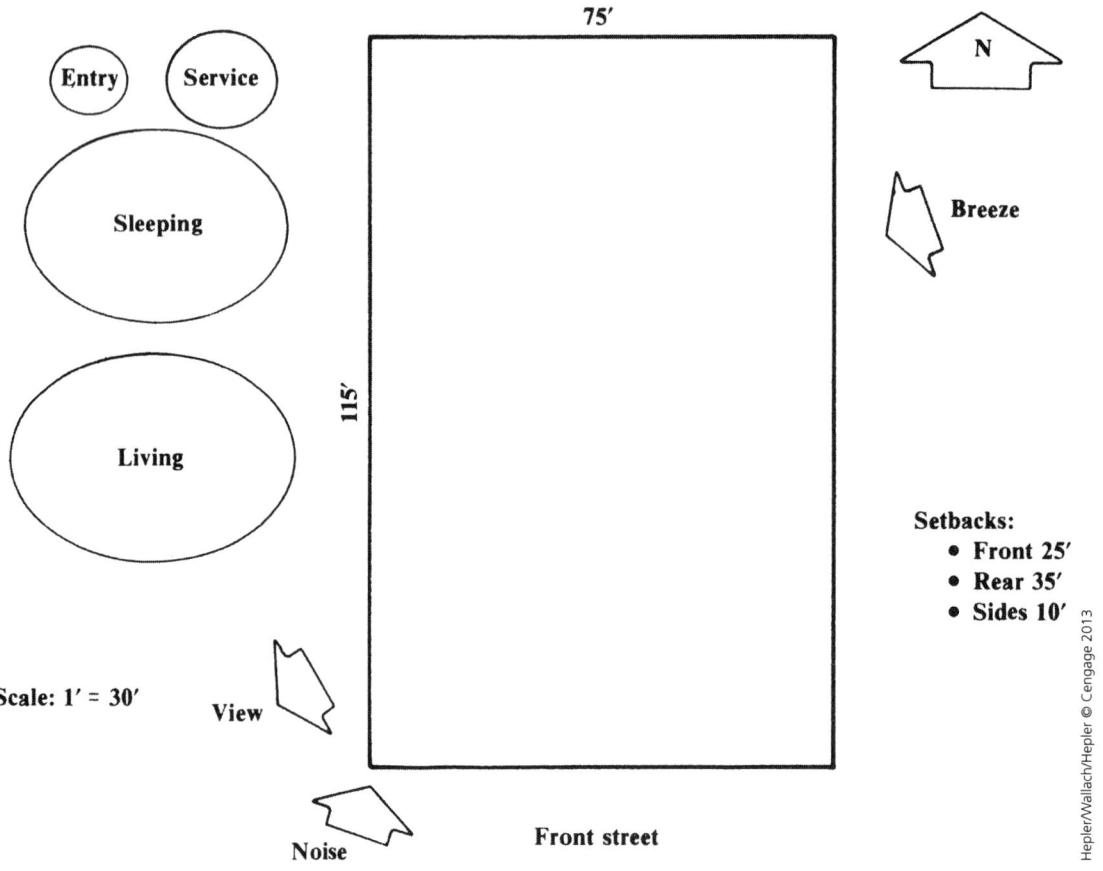

Arrange the bubbles for an efficient orientation on the irregular-shaped lot. Use the scame scale and setbacks as above.

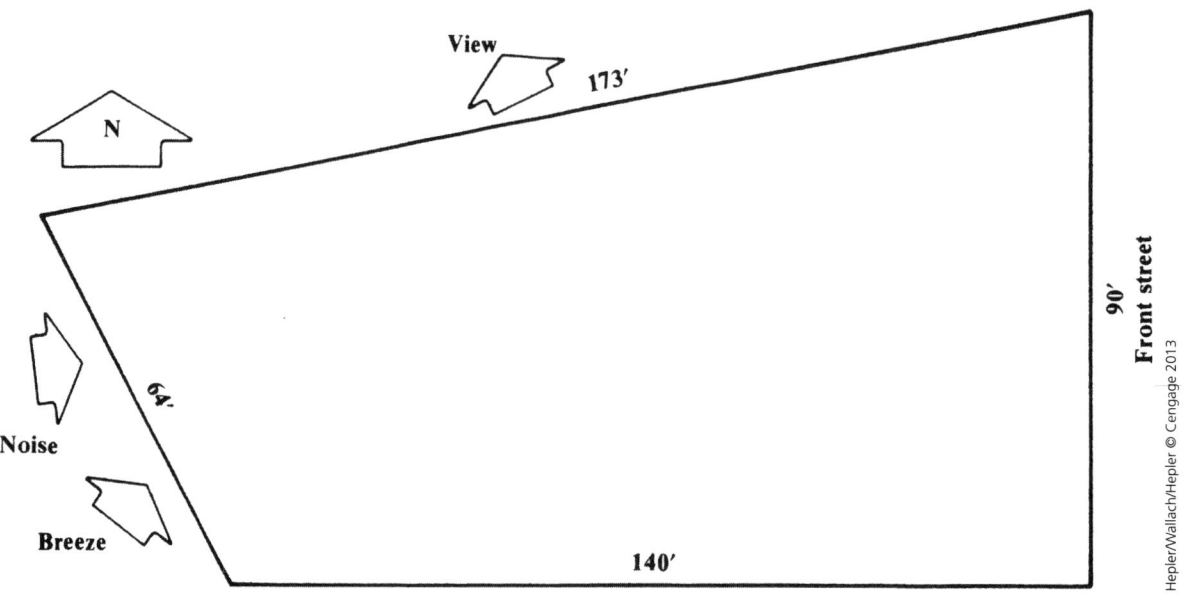

TITLE	DRAFTER	SCALE	DATE	DWG NO.

EXERCISE 6-4

Figure the roof overhang for the following latitudes.

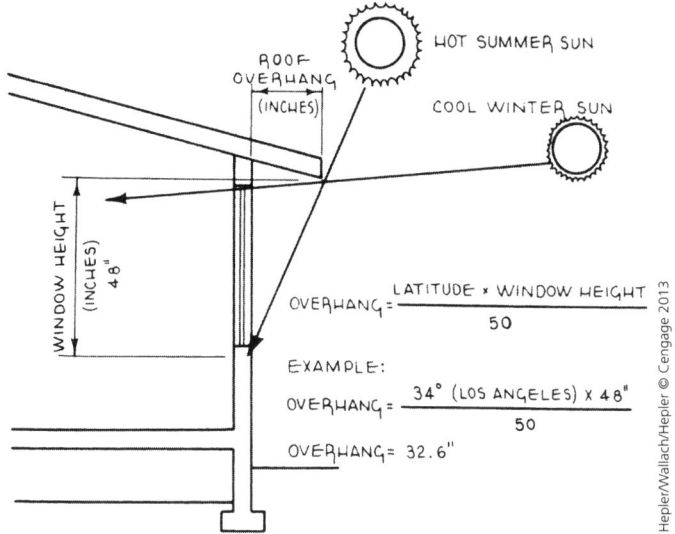

1. 15° _____

2. 30° _____

3. 45° _____

4. 60° _____

5. 72° _____

Trace templates for the house and garage. Orient and place within the setbacks of 20′ front, 30′ rear, and 6′ side setbacks. Explain your choice of orientation.

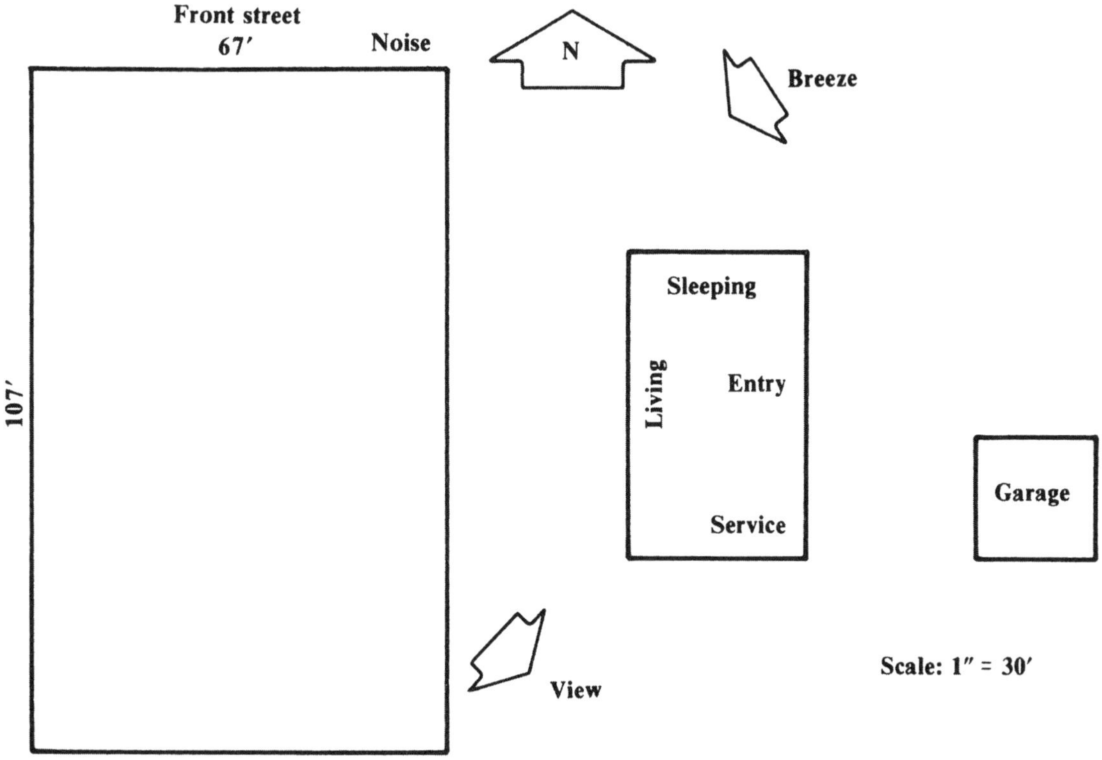

TITLE	DRAFTER	SCALE	DATE	DWG NO.

EXERCISE 7-1

Add walls, fireplace, and built-in components to the living and dining area on this plan.

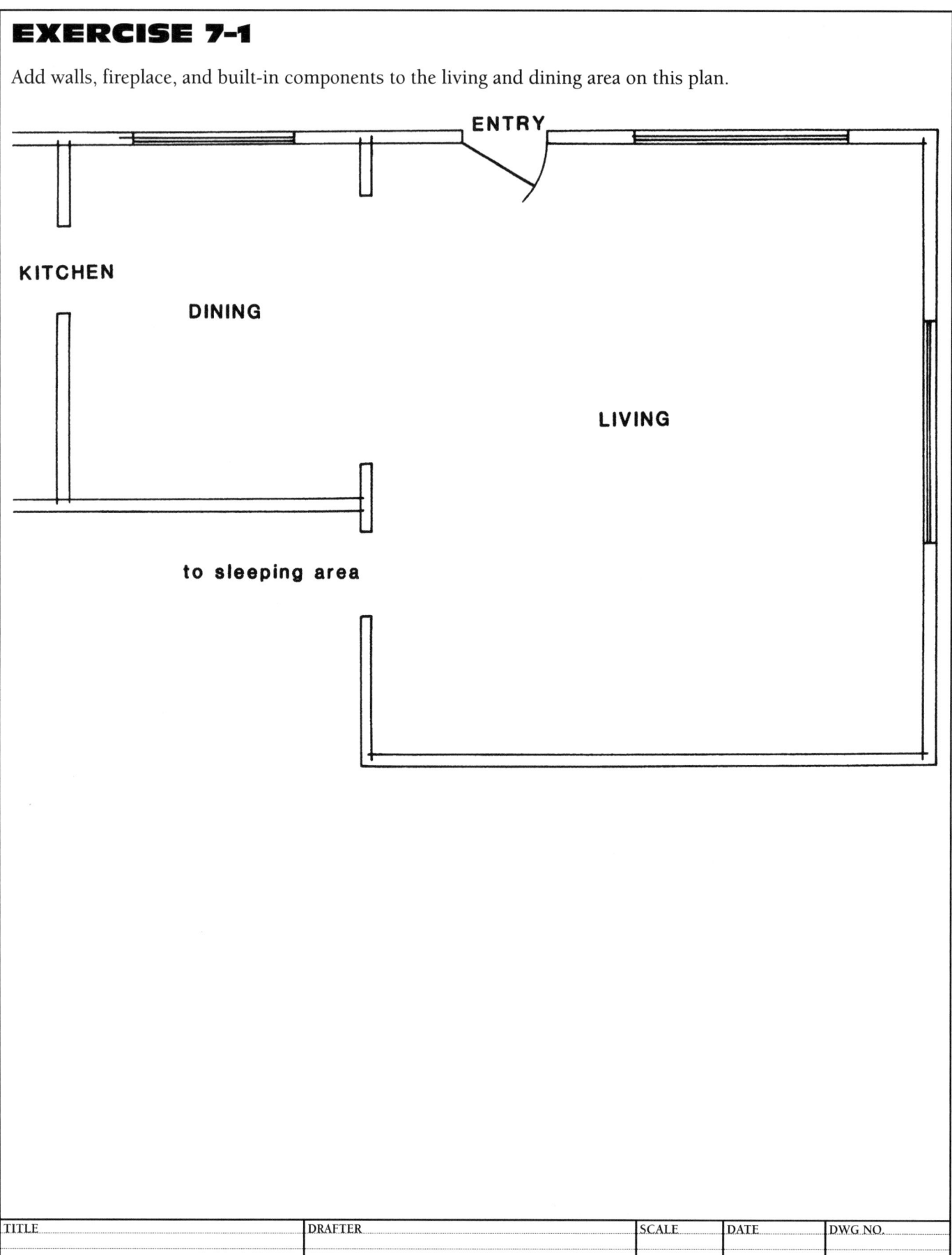

EXERCISE 7-2

In the space provided, design a living area that has a living room, dining room, and den. Use a scale of 1/8″ = 1′–0″.

TITLE	DRAFTER	SCALE	DATE	DWG NO.

EXERCISE 8-1

Sketch the following additions to this plan: (1) quiet patio for sleeping area; (2) children's play patio; (3) dining patio; and (4) outdoor service area.

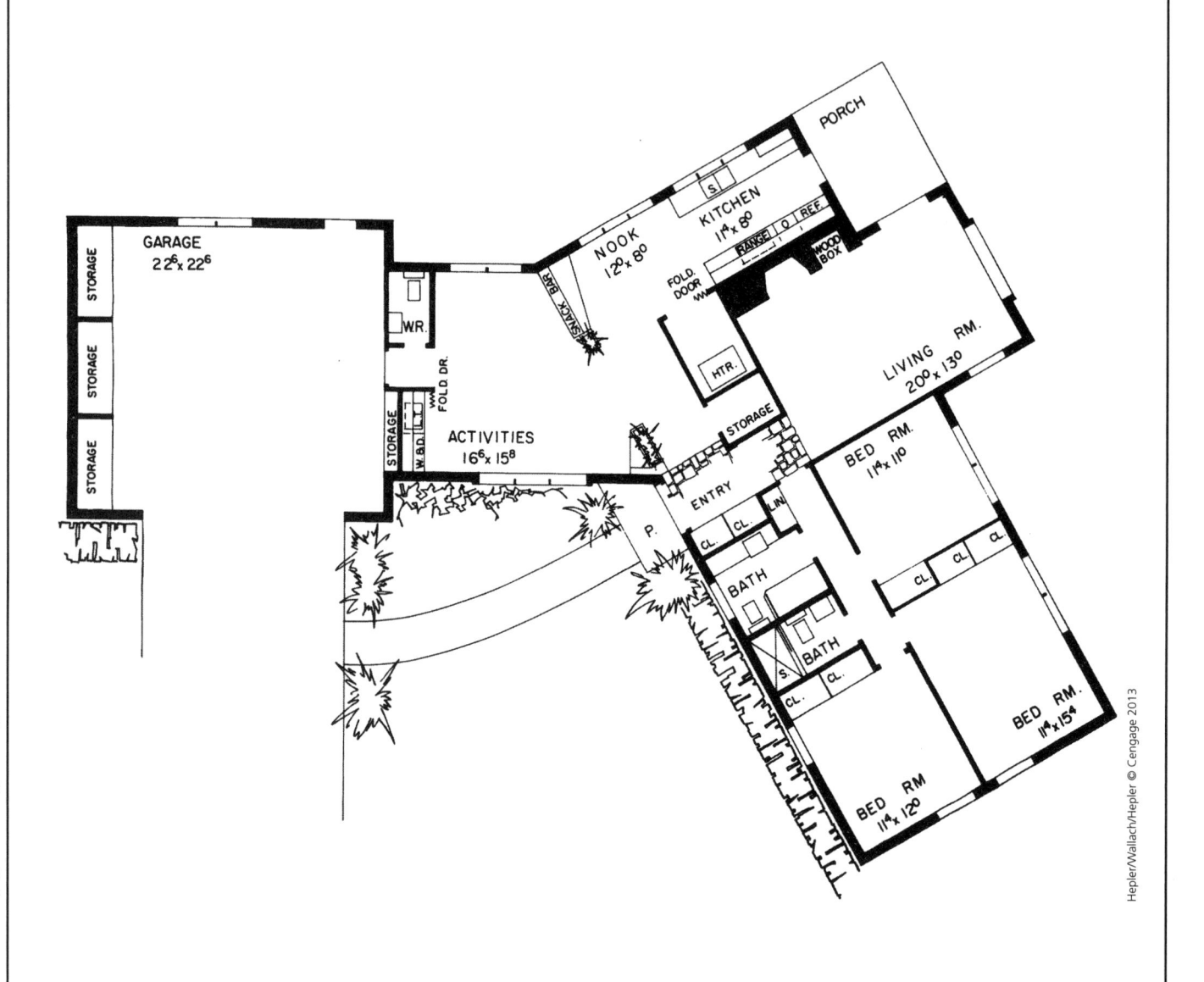

TITLE	DRAFTER	SCALE	DATE	DWG NO.

EXERCISE 8-2

Sketch the following additions to this plan: (1) quiet patio for sleeping area; (2) children's play patio; (3) dining patio; and (4) outdoor service area.

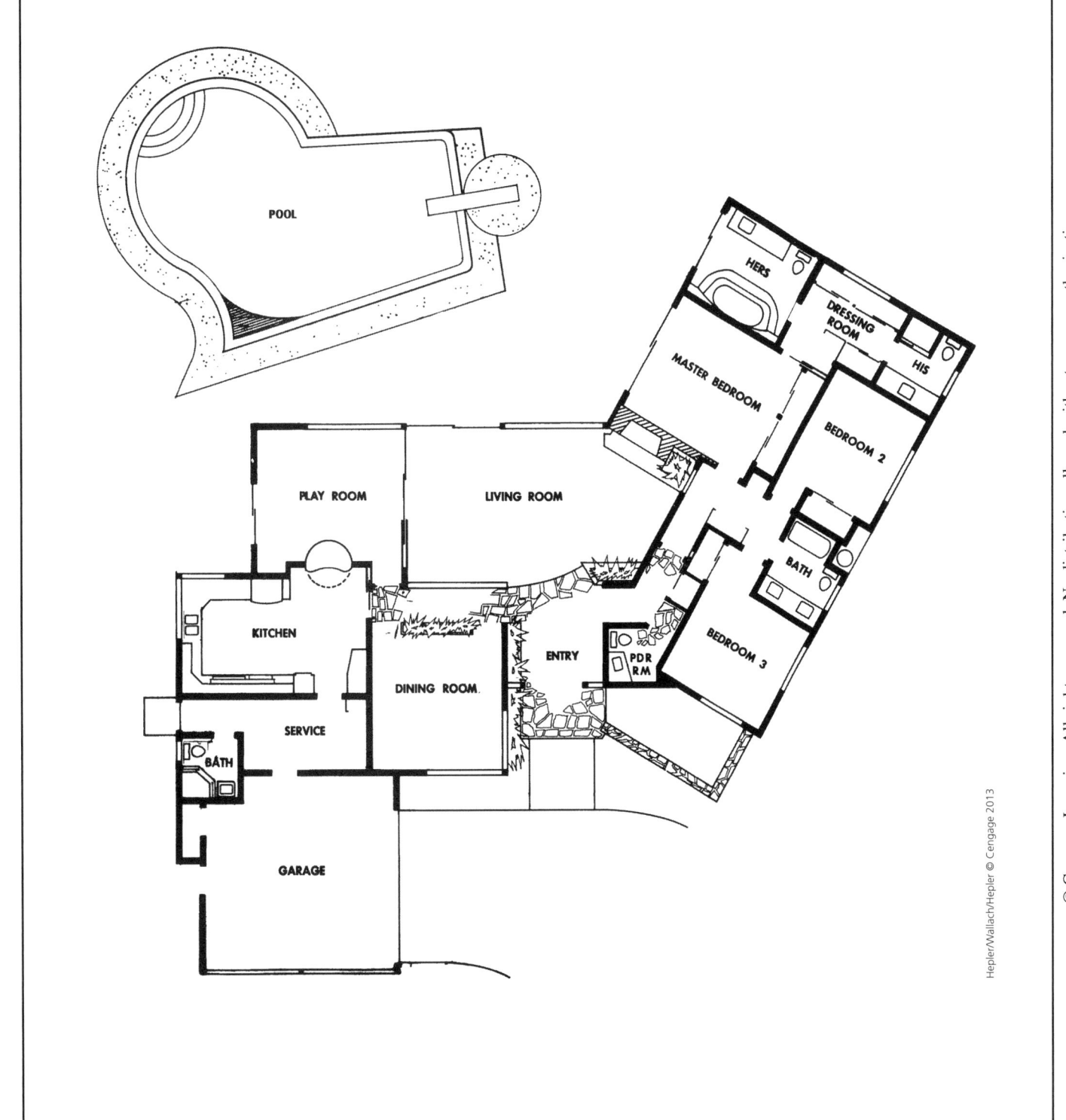

TITLE	DRAFTER	SCALE	DATE	DWG NO.

EXERCISE 8-3

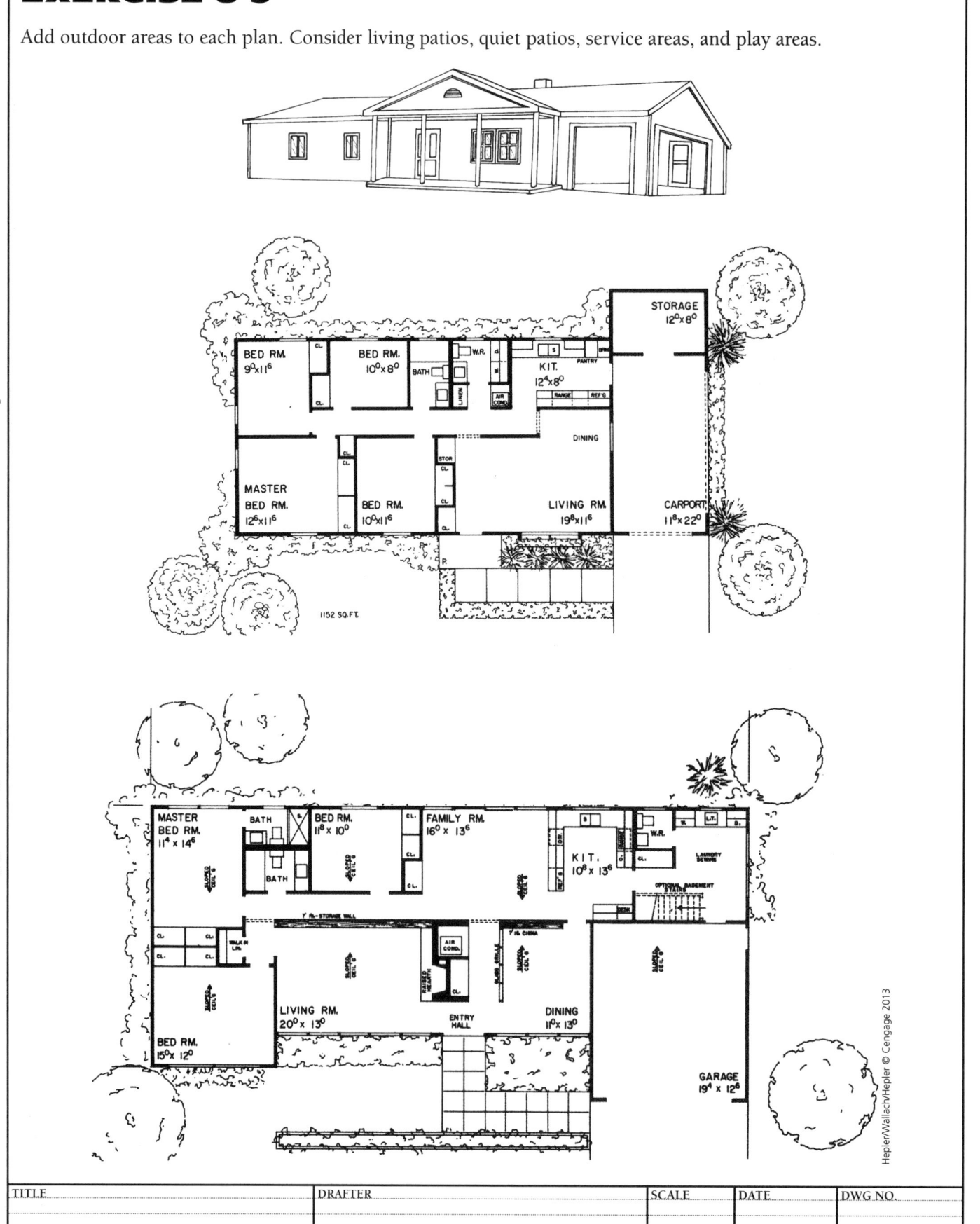

Add outdoor areas to each plan. Consider living patios, quiet patios, service areas, and play areas.

TITLE	DRAFTER	SCALE	DATE	DWG NO.

EXERCISE 9-1

Study the floor plan and write the approximate distances from one area to another on the table. Then rate the traffic flow for the areas listed by making a checkmark in the appropriate column.

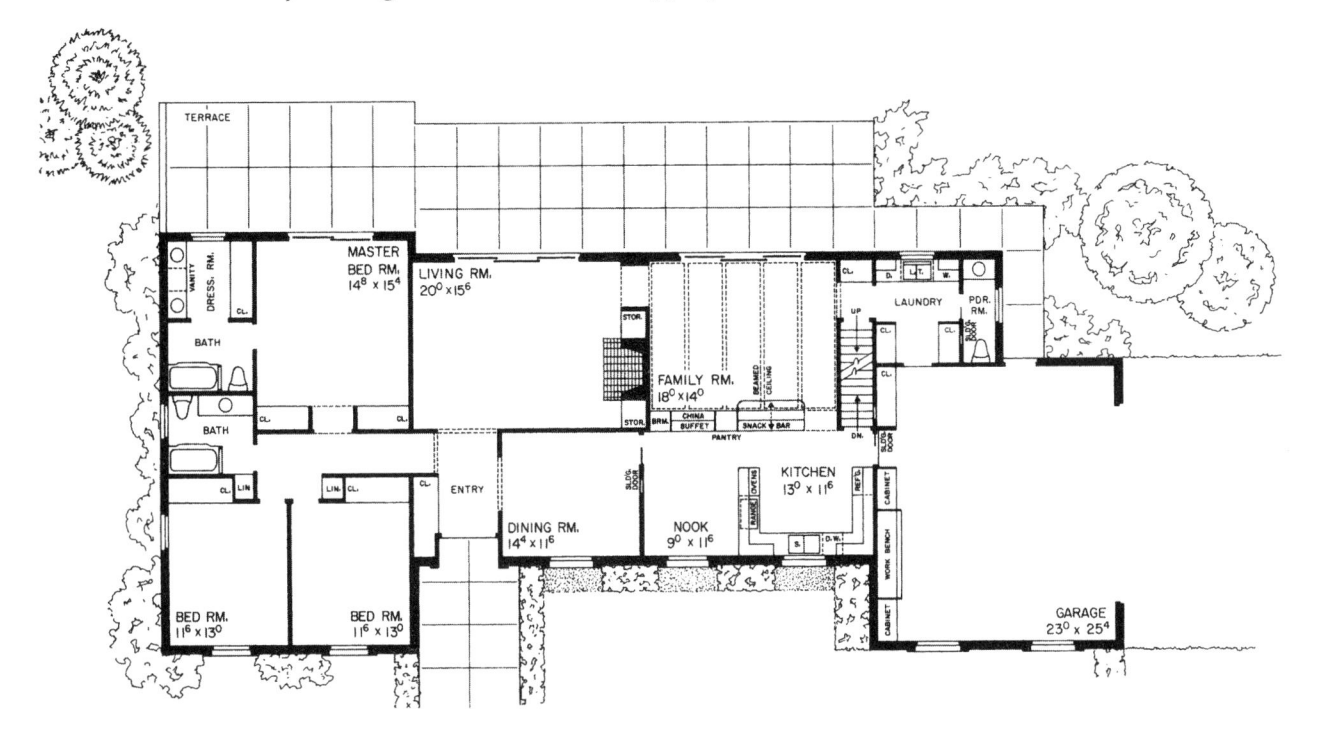

	APPROXIMATE DISTANCE	EXCELLENT	FAIR	POOR
SERVICE CIRCULATION				
Service door to kitchen				
Service door to living area				
Service door to bathroom				
Service door to bedrooms				
Kitchen to bathrooms				
Paths around work triangle				
GUEST'S CIRCULATION				
Front door to closet				
Front door to living room				
Front door to bathroom				
Front door to outdoor living				
Living room to bathroom				
RESIDENTS' CIRCULATION				
Front door to kitchen				
Kitchen to bedrooms				
Kitchen to children's play area				
Kitchen to children's sleeping area				
Bedrooms to bathrooms				
Outdoor living area to bathroom				
Living room to outdoor living area				

TITLE	DRAFTER	SCALE	DATE	DWG NO.

EXERCISE 9-2

Study both floor plans. Then, using a colored pencil, draw the main traffic patterns on the plans.

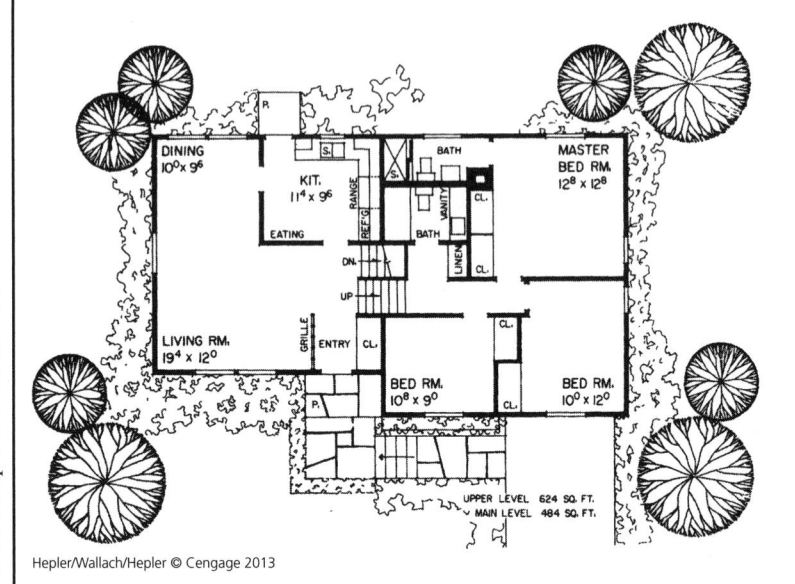

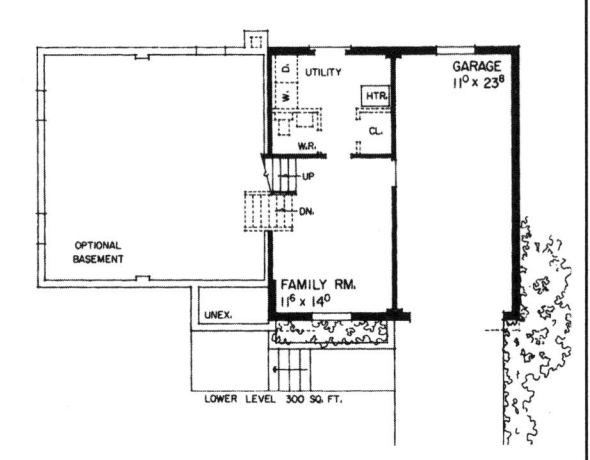

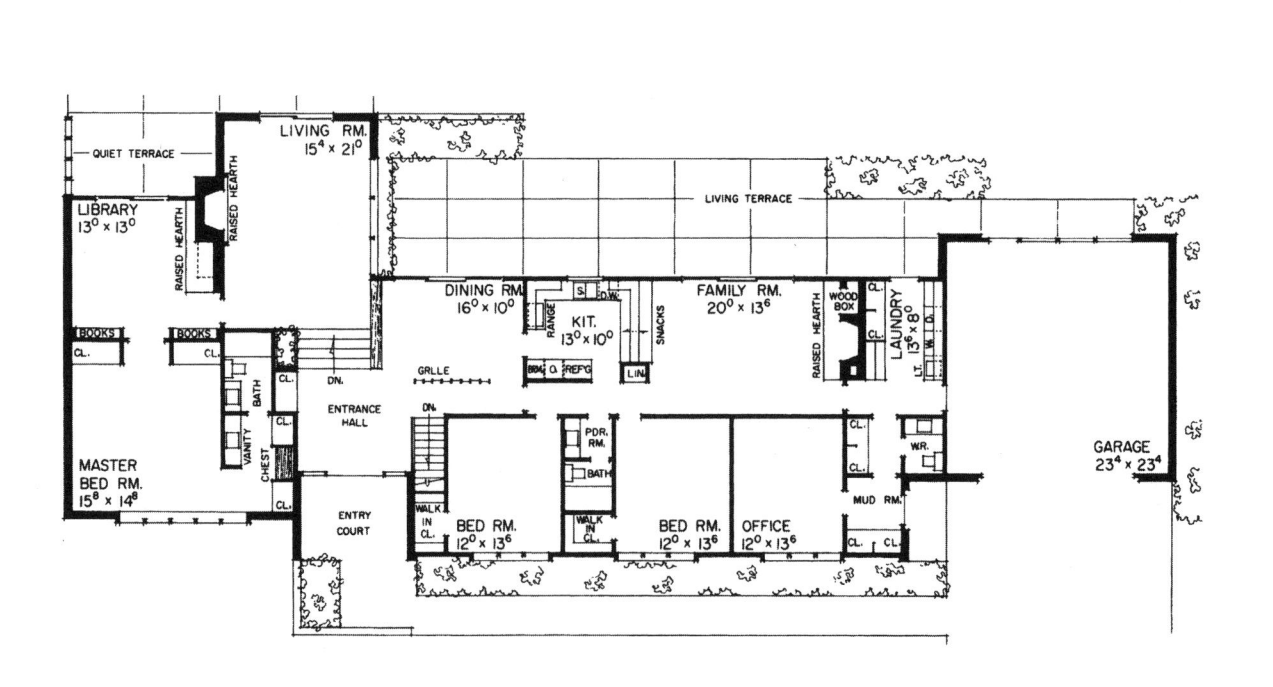

TITLE	DRAFTER	SCALE	DATE	DWG NO.

EXERCISE 10-1

Make a drawing of a simple kitchen (no more than 200 sq. ft.) of your own design. As you work, remember the guidelines for designing kitchens:

- Floor cabinets 24" to 26" deep
- Wall cabinets 12" to 15" deep
- Sink located between the refrigerator and range
- Range located near the dining area
- Perimeter of work triangle between 12' and 21'
- Counter work space available on both sides of appliances
- Window over the sink
- At least 4' of floor space between opposing floor cabinets
- Adequate lighting for all work areas
- Adequate electrical outlets for all work areas
- No traffic flow through the work triangle
- Refrigerator and food storage adjacent to service entry

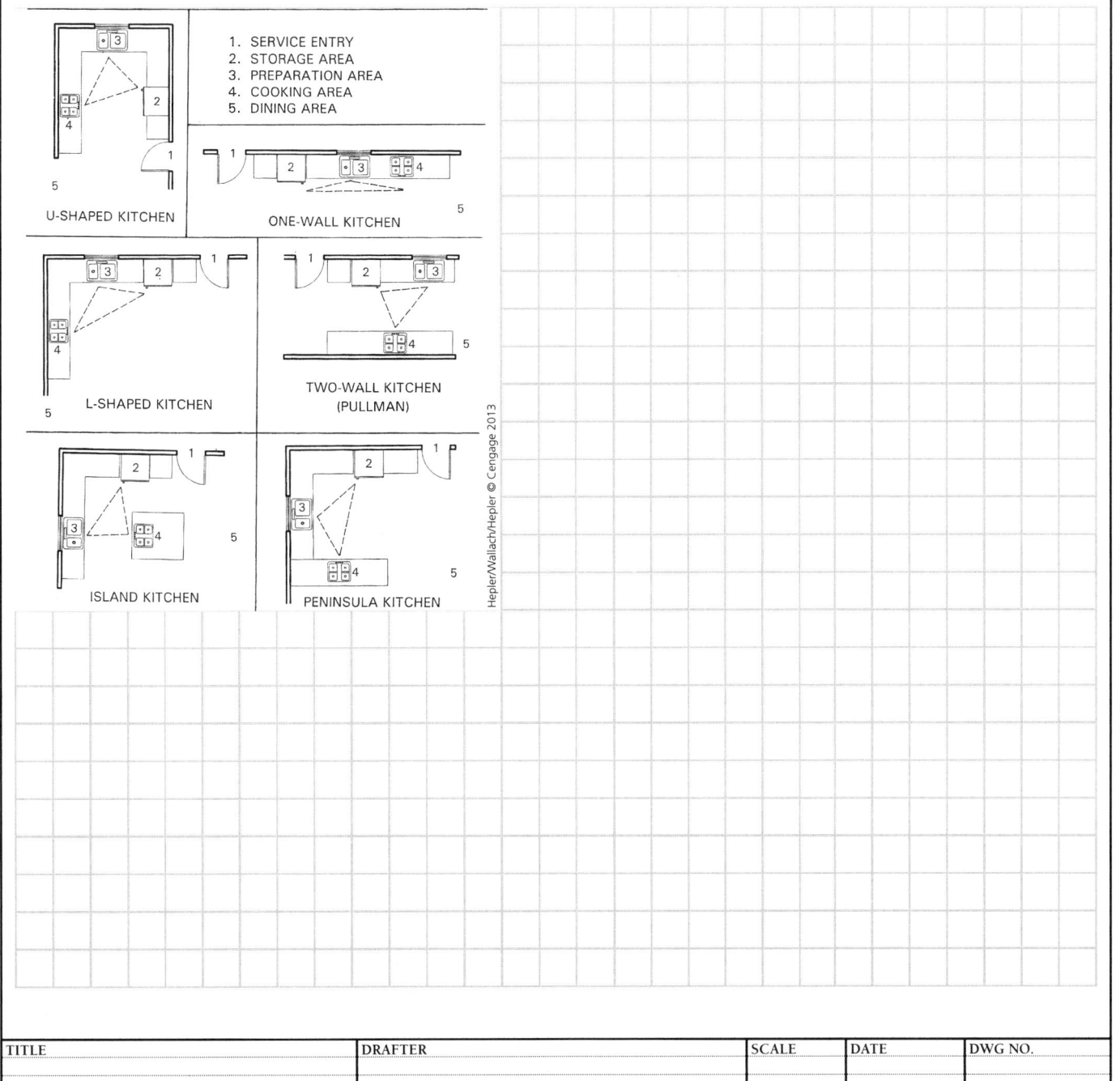

1. SERVICE ENTRY
2. STORAGE AREA
3. PREPARATION AREA
4. COOKING AREA
5. DINING AREA

U-SHAPED KITCHEN

ONE-WALL KITCHEN

L-SHAPED KITCHEN

TWO-WALL KITCHEN (PULLMAN)

ISLAND KITCHEN

PENINSULA KITCHEN

TITLE	DRAFTER	SCALE	DATE	DWG NO.

EXERCISE 10-2

In the space below, design a full-service kitchen.

| TITLE | DRAFTER | SCALE | DATE | DWG NO. |

EXERCISE 10-3

Design and draw an alternative family kitchen below using the same space as shown in the plan at the top. Use a scale of 1/4" = 1'–0".

SCALE: 1/4"=1'-0"

TITLE	DRAFTER	SCALE	DATE	DWG NO.

EXERCISE 10-4

Design and draw an alternative peninsula or island kitchen below at a scale of 1/2" = 1'–0".

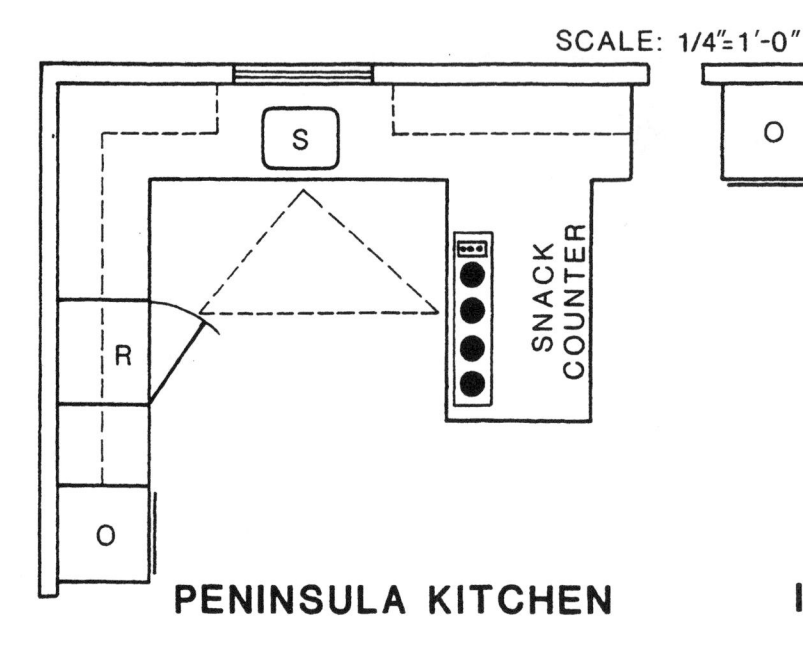

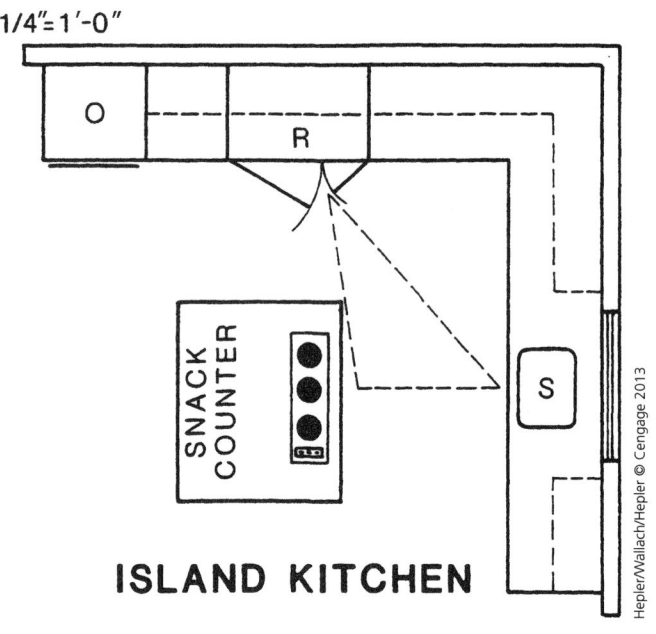

SCALE: 1/4"=1'-0"

PENINSULA KITCHEN

ISLAND KITCHEN

TITLE	DRAFTER	SCALE	DATE	DWG NO.

EXERCISE 10-5

Design and draw an alternative corridor or L-shape kitchen below at a sale of 1/2″ = 1′–0″.

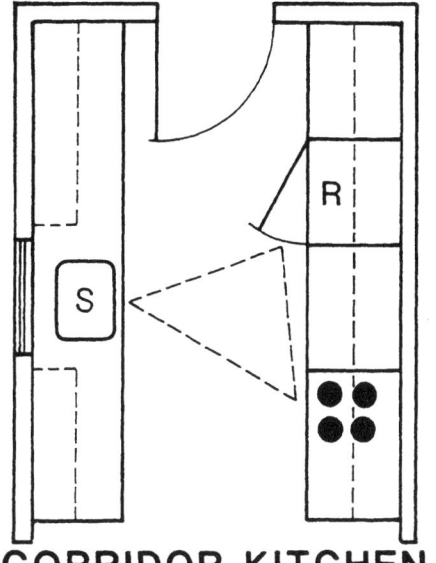

CORRIDOR KITCHEN

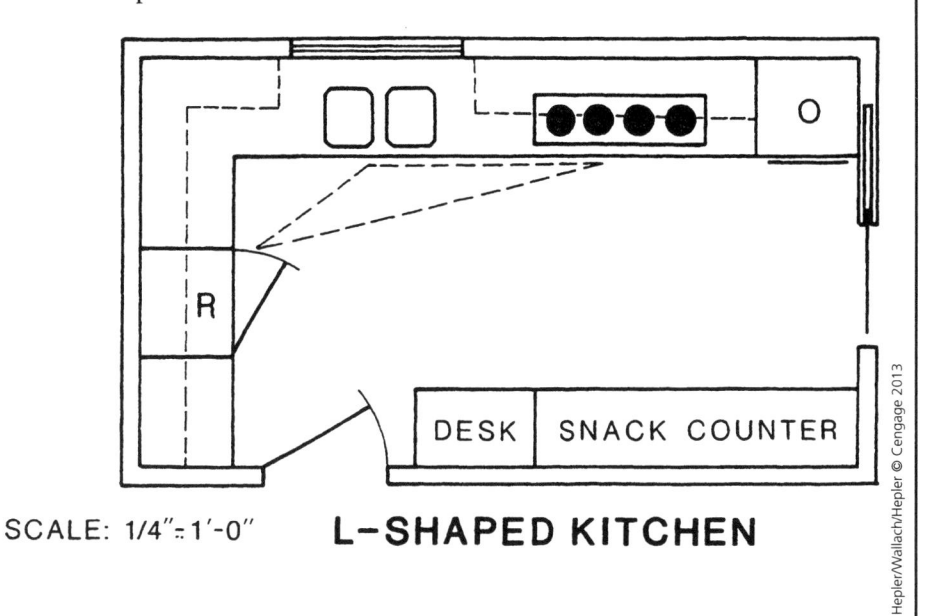

SCALE: 1/4″ = 1′-0″

L-SHAPED KITCHEN

TITLE	DRAFTER	SCALE	DATE	DWG NO.

EXERCISE 10-6

Design and draw a U-shaped kitchen with a breakfast area. Use a scale of 1/2" = 1'–0".

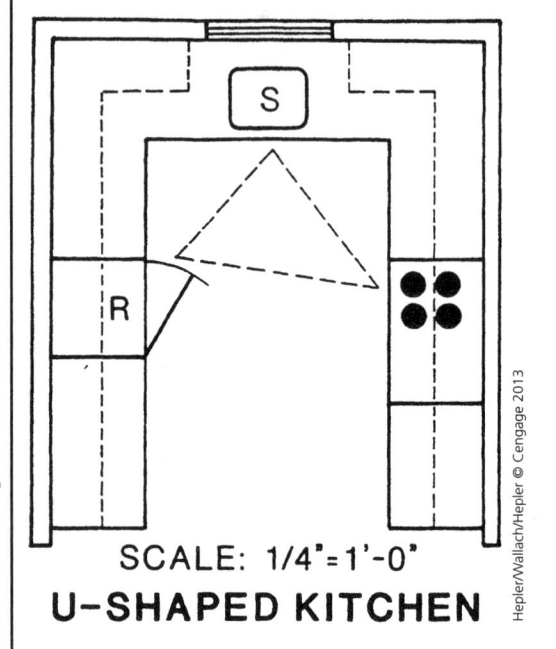

General Kitchen Design Rules
1. Floor cabinets 24" to 26" deep
2. Wall cabinets 12" to 15" deep
3. Sink located between the refrigerator and range
4. Range located near the dining area
5. Perimeter of work triangle between 12' and 21'
6. Counter work space available on both sides of appliances
7. Window over the sink
8. At least 4' of floor space between opposing floor cabinets
9. Adequate lightning for all work areas
10. Adequate electrical outlets for all work areas
11. No traffic flow through the work triangle
12. Refrigerator and food storage adjacent to service entry

SCALE: 1/4"=1'-0"
U-SHAPED KITCHEN

TITLE	DRAFTER	SCALE	DATE	DWG NO.

EXERCISE 11-1, PART A

Add closets, storage, appliances, fixtures, counters, and cabinets to the service area shown below.

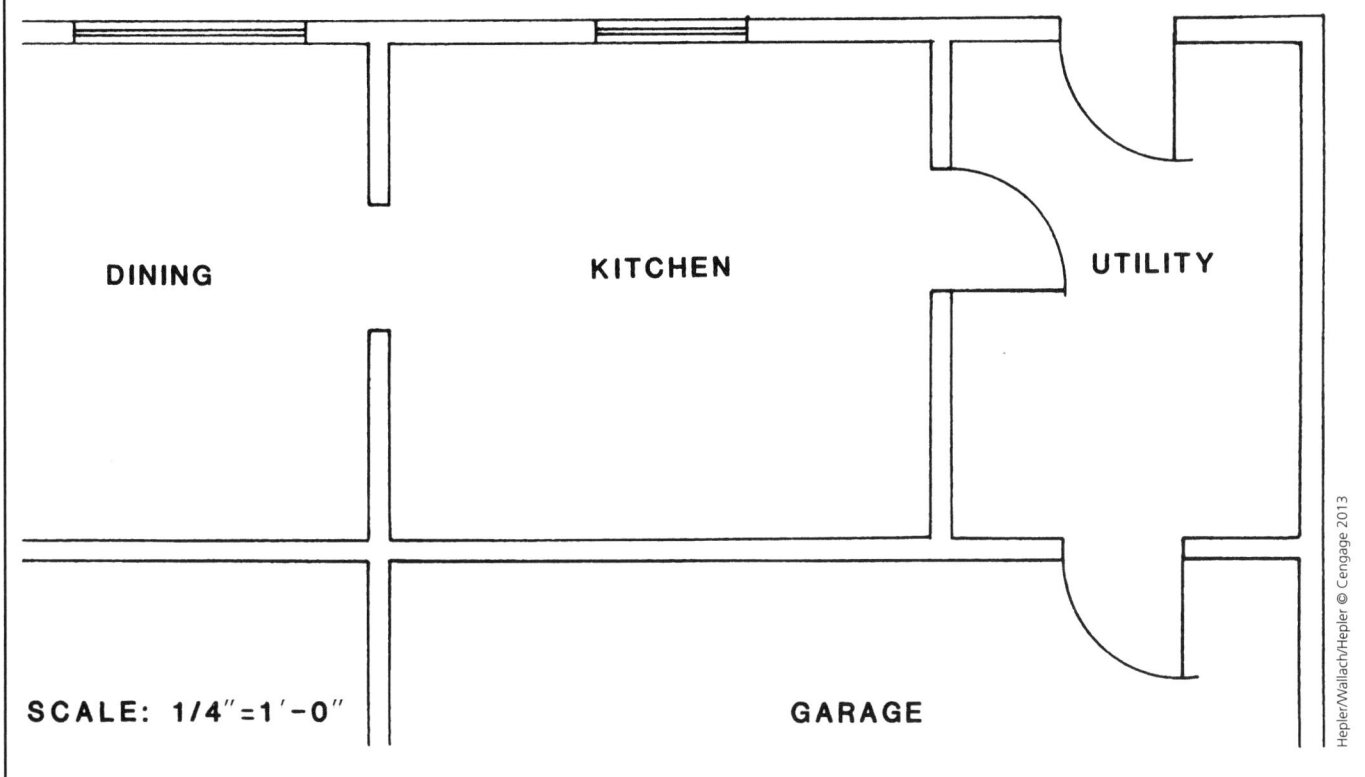

DINING   KITCHEN   UTILITY

SCALE: 1/4"=1'-0"   GARAGE

EXERCISE 11-1, PART B

Using the grid below, design a full-service utility room that measures 12" × 10'.

| TITLE | DRAFTER | SCALE | DATE | DWG NO. |

EXERCISE 11-2

Using the grid below, design a two-car garage that measures 20' by 30' and that has a full utility room.

TITLE	DRAFTER	SCALE	DATE	DWG NO.

EXERCISE 12-1

Complete the plan for each room by adding doors, windows, and fixtures. As you work, remember the guidelines for designing bathrooms:

- Water closet clearances—15″ minimum for sides and 18″ for front
- Tub clearance—30″ for sides
- Lavatory clearance—18″ for sides
- Standard tub lengths, 5′–0″—5′–6,″ and 6′–0″
- Typical bathroom door width—2′–2″; for wheelchairs, 2′–9″

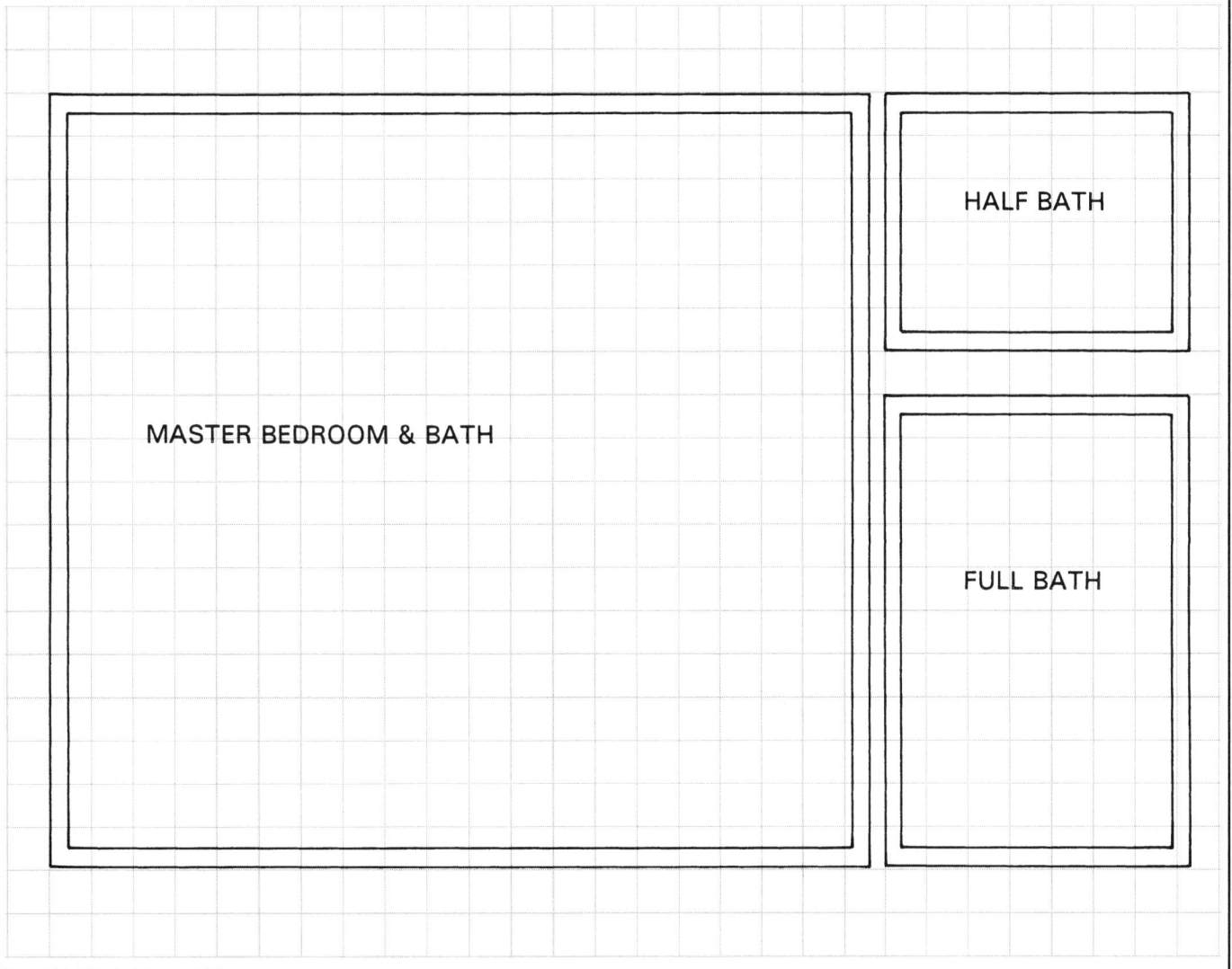

MASTER BEDROOM & BATH

HALF BATH

FULL BATH

Hepler/Wallach/Hepler © Cengage 2013

| TITLE | DRAFTER | SCALE | DATE | DWG NO. |

EXERCISE 12-2 To the bedroom areas shown below, add a bathroom, dressing area, windows, doors, closets, storage, and furniture outlines.

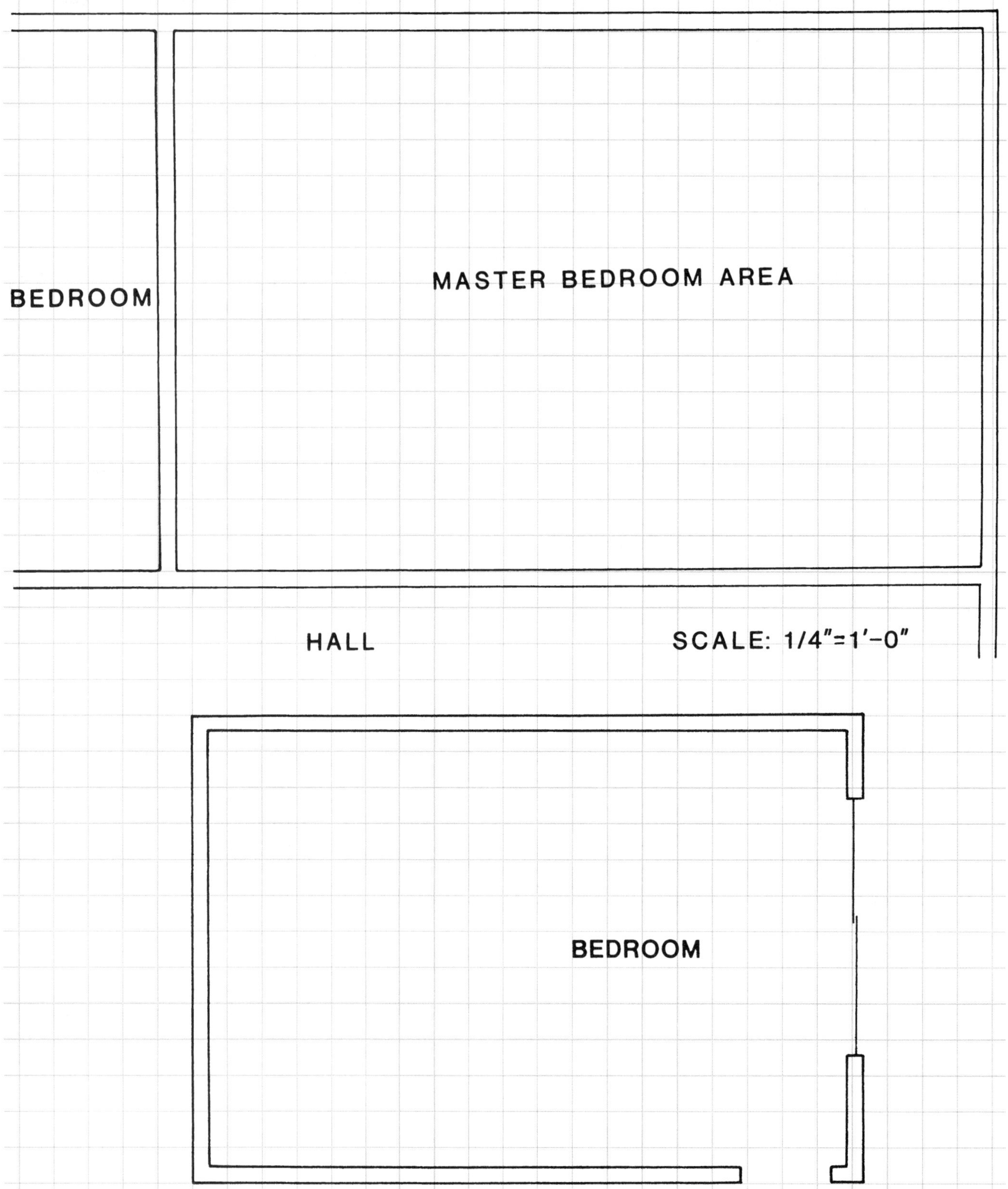

BEDROOM

MASTER BEDROOM AREA

HALL   SCALE: 1/4"=1'-0"

BEDROOM

TITLE	DRAFTER	SCALE	DATE	DWG NO.

EXERCISE 12-3

Design a bathroom for an 8′ × 14′ space. Use a scale of 1/4″ = 1′–0″.

TYPICAL BATHROOM DESIGNS AND FIXTURE CLEARANCES

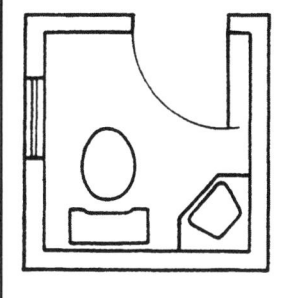

HALF BATH

SCALE: 1/4″=1′-0″

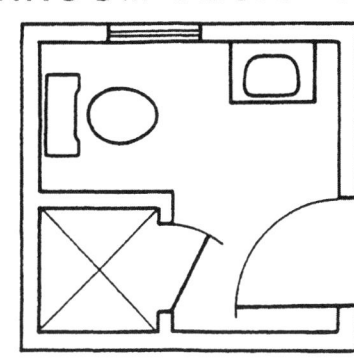

3/4 BATH

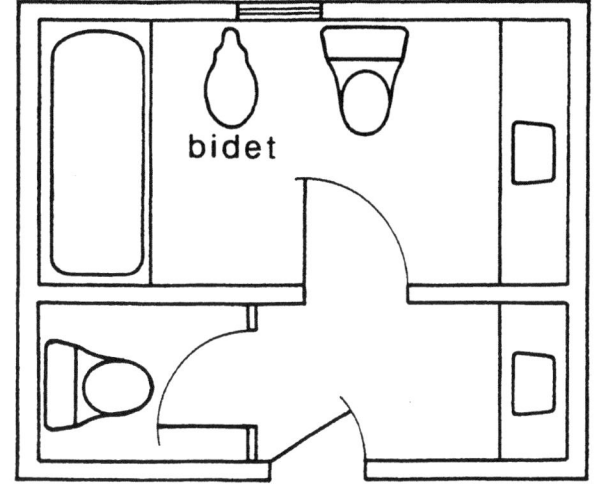

COMPARTMENTED BATH

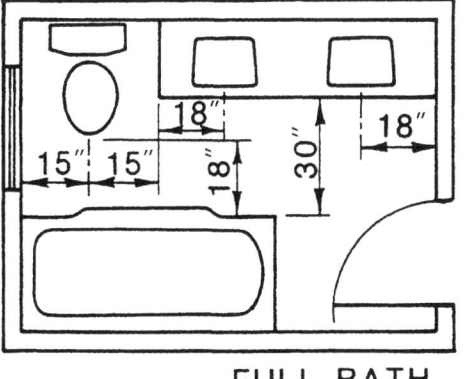

bidet

FULL BATH

BATHROOM FIXTURE CLEARANCES & DESIGN FACTORS

1. Water closet clearances - 15″ minimum for sides and 18″ for front
2. Tub clearance - 30″ for sides
3. Lavatory clearance - 18″ for sides
4. Standard tub lengths - 5′-0″, 5′-6″, and 6′-0″
5. Typical bathroom door width - 2′-2″; for wheelchairs, 2′-9″

TITLE	DRAFTER	SCALE	DATE	DWG NO.

EXERCISE 12-4

Design a different master bedroom suite using the same space as in this plan using a scale of 1/4" = 1'–0."

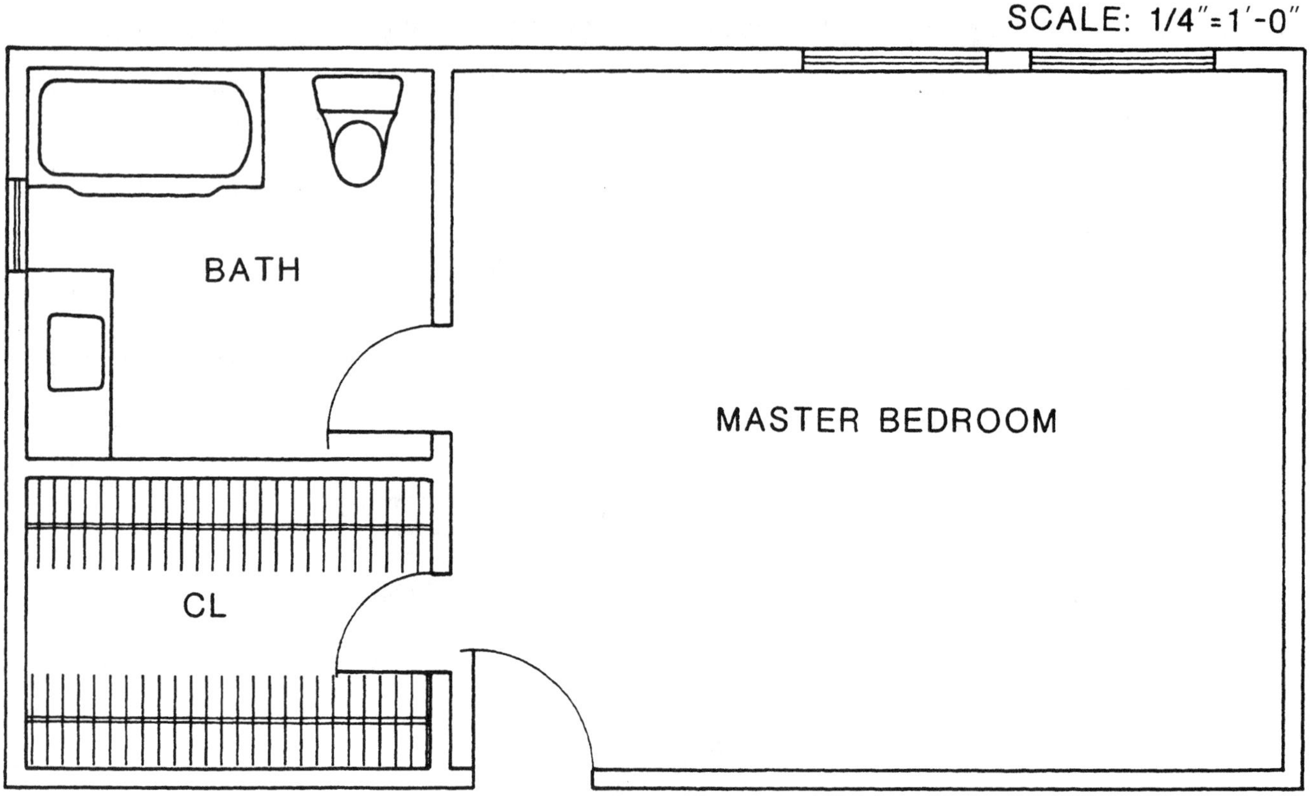

SCALE: 1/4"=1'-0"

BATH

MASTER BEDROOM

CL

TITLE	DRAFTER	SCALE	DATE	DWG NO.

EXERCISE 13-1, PART A

Complete the plot plan below, using a combination of single- and multiple-family dwellings in order to house 20 families. Include each structure on the plan and add driveways, parking and recreational facilities, and landscape features.

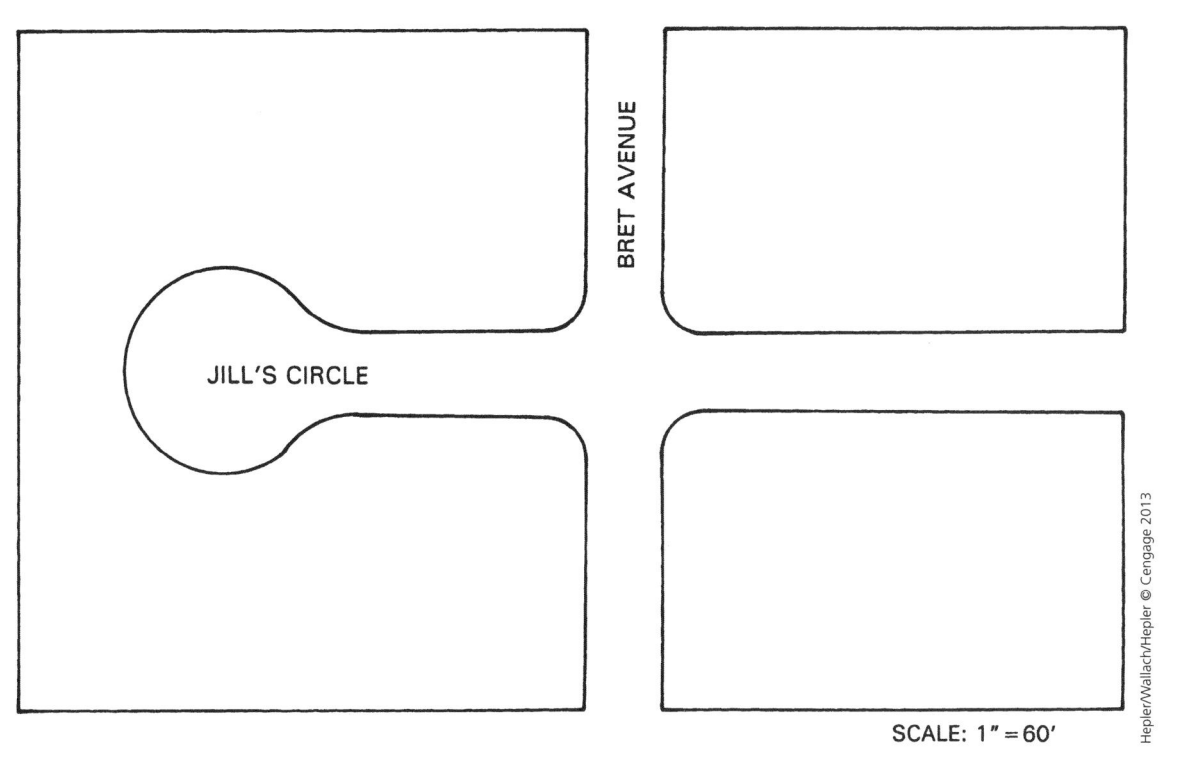

JILL'S CIRCLE

BRET AVENUE

SCALE: 1" = 60'

EXERCISE 13-1, PART B

Add facilities to this lot that will allow maximum use of the outdoors.

DRIVEWAY

WINDS

N

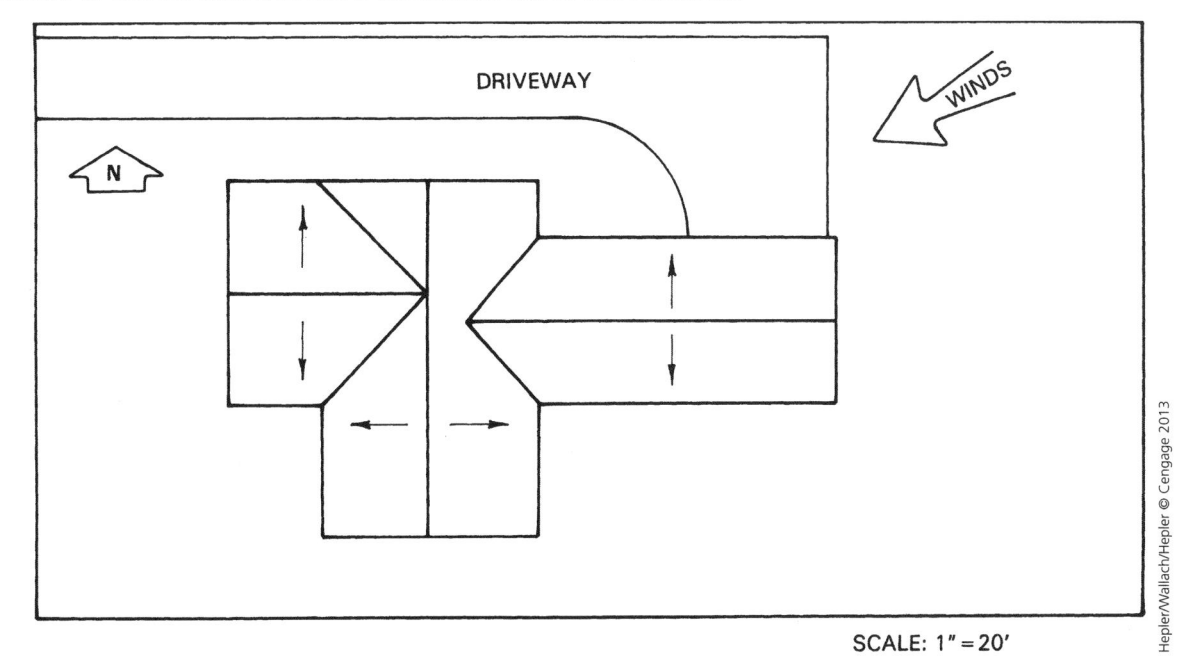

SCALE: 1" = 20'

| TITLE | DRAFTER | SCALE | DATE | DWG NO. |

EXERCISE 13-2 Develop this parcel of land into a subdivision. First, divide the area into eight lots. Each lot should have a single-family house measuring 1,500 sq. ft. (An example is shown below.) Orient each home for maximum variety, efficiency, and privacy. Include parking, streets, and landscape features.

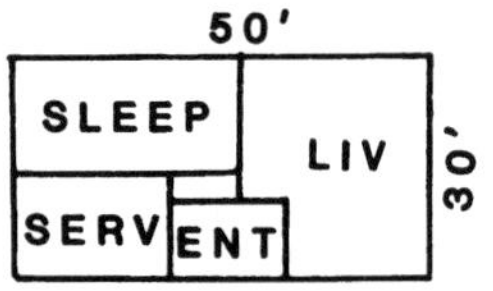

SCALE: 1"=40'

TITLE	DRAFTER	SCALE	DATE	DWG NO.

EXERCISE 13 Site Development Plans ■ 55

EXERCISE 13-3
Develop profiles X–X and Y–Y, using contour intervals of 1'–0". Show the elevation of the building and the depth of the septic fields.

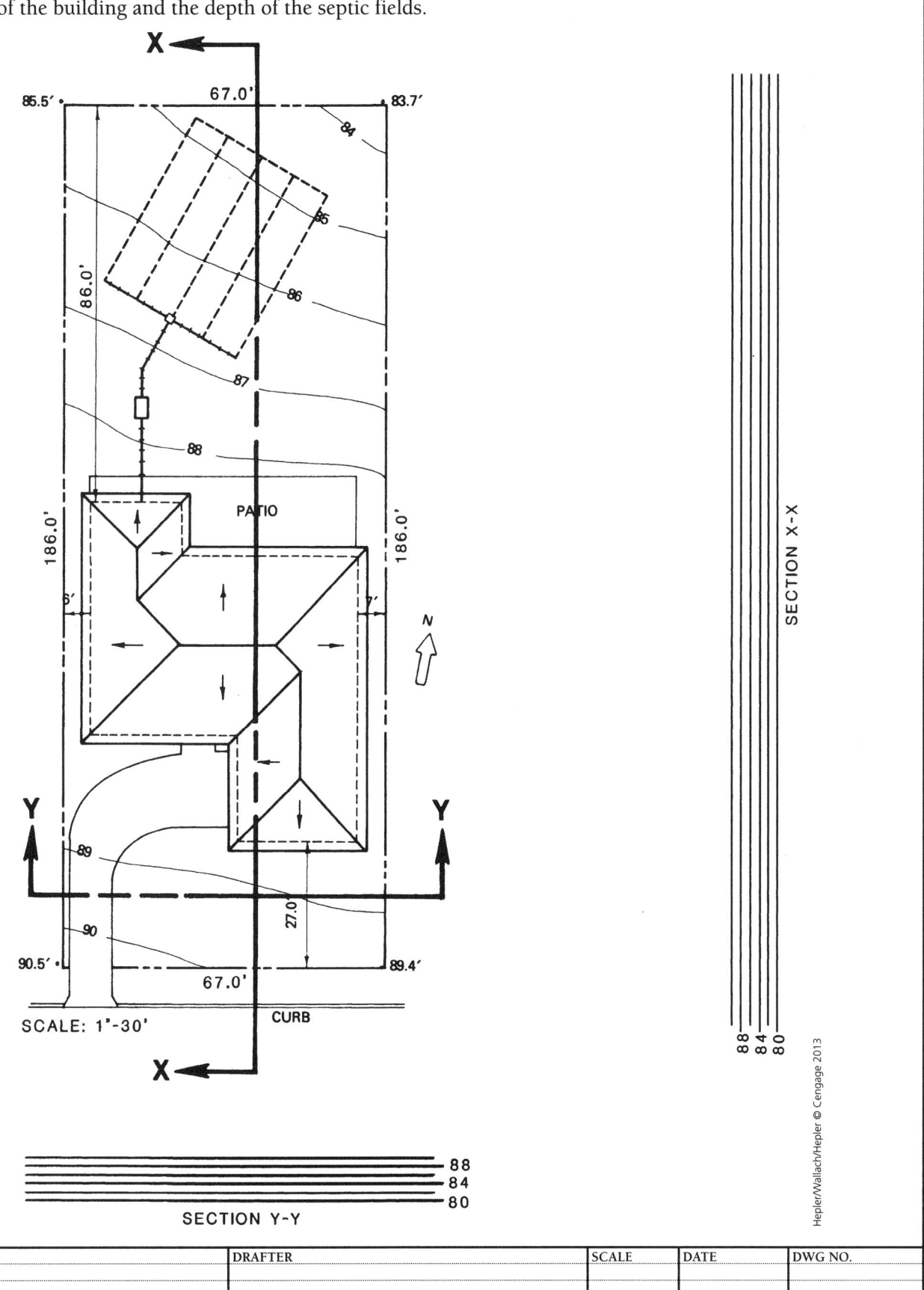

SCALE: 1"-30'

SECTION X-X

SECTION Y-Y

TITLE	DRAFTER	SCALE	DATE	DWG NO.

EXERCISE 13-4

Position a house on the lot. Then adjust contour lines to indicate final grading. Develop profiles X–X, Y–Y, and Z–Z to match your design. If necessary, move cutting plane lines so that they intersect your house.

TITLE	DRAFTER	SCALE	DATE	DWG NO.

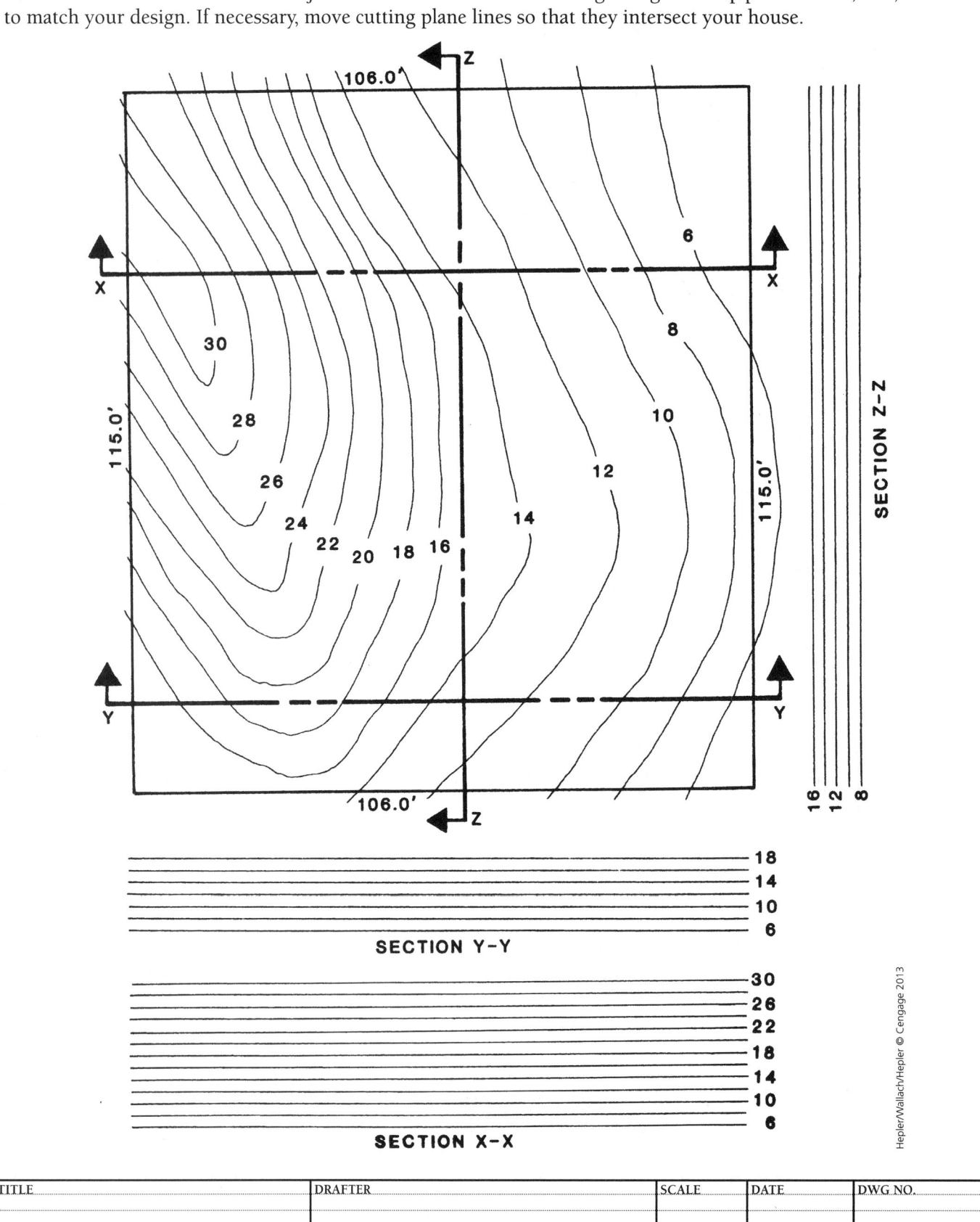

EXERCISE 13-5

Design a house for this family on the lot shown below. Information that will help you with your design is listed below.

- Two adults
- One teenage boy
- One middle school girl
- One small dog
- Household income is $100,000

The zoning laws for this lot are:

- 25' front setback
- 20' rear setback
- 6' side setback
- Maximum coverage of home should not exceed 50% of the lot.

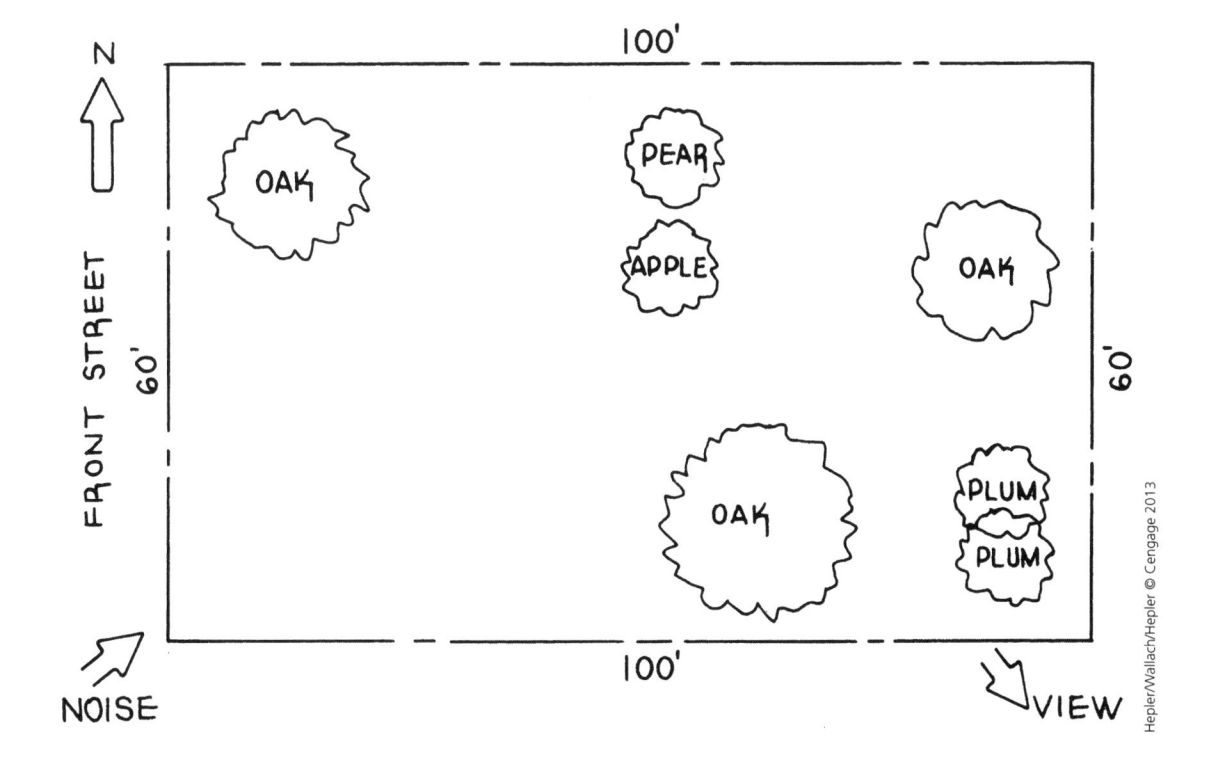

| TITLE | DRAFTER | SCALE | DATE | DWG NO. |

EXERCISE 13-6

Complete the plot plan using a scale of 3/16″ = 1′–0″. The setbacks are 5′ for the sides and 20′ for the front and rear. Orient the house on the site and add parking, driveways, walks, landscaping, and dimensions.

FRONT STREET

N

TITLE	DRAFTER	SCALE	DATE	DWG NO.

EXERCISE 14-1 Following the exercises in this workbook is a series of templates. Find T–1: Floor Plan Rooms on page 180 and cut out the templates. Use them to design a house. Sketch your design on the grid below. *Note:* Scale for the templates is 1/8″ = 1′–0″; the grid scale is 1/4″ = 1′–0″.

TITLE	DRAFTER	SCALE	DATE	DWG NO.

EXERCISE 14-2 Following the exercises in this workbook is a series of templates. Find T–2: Living Area and T–3: Bedroom and Service Areas on pages 181 and 182 and cut the templates out. Use them to create functional furniture arrangements on the plans below.

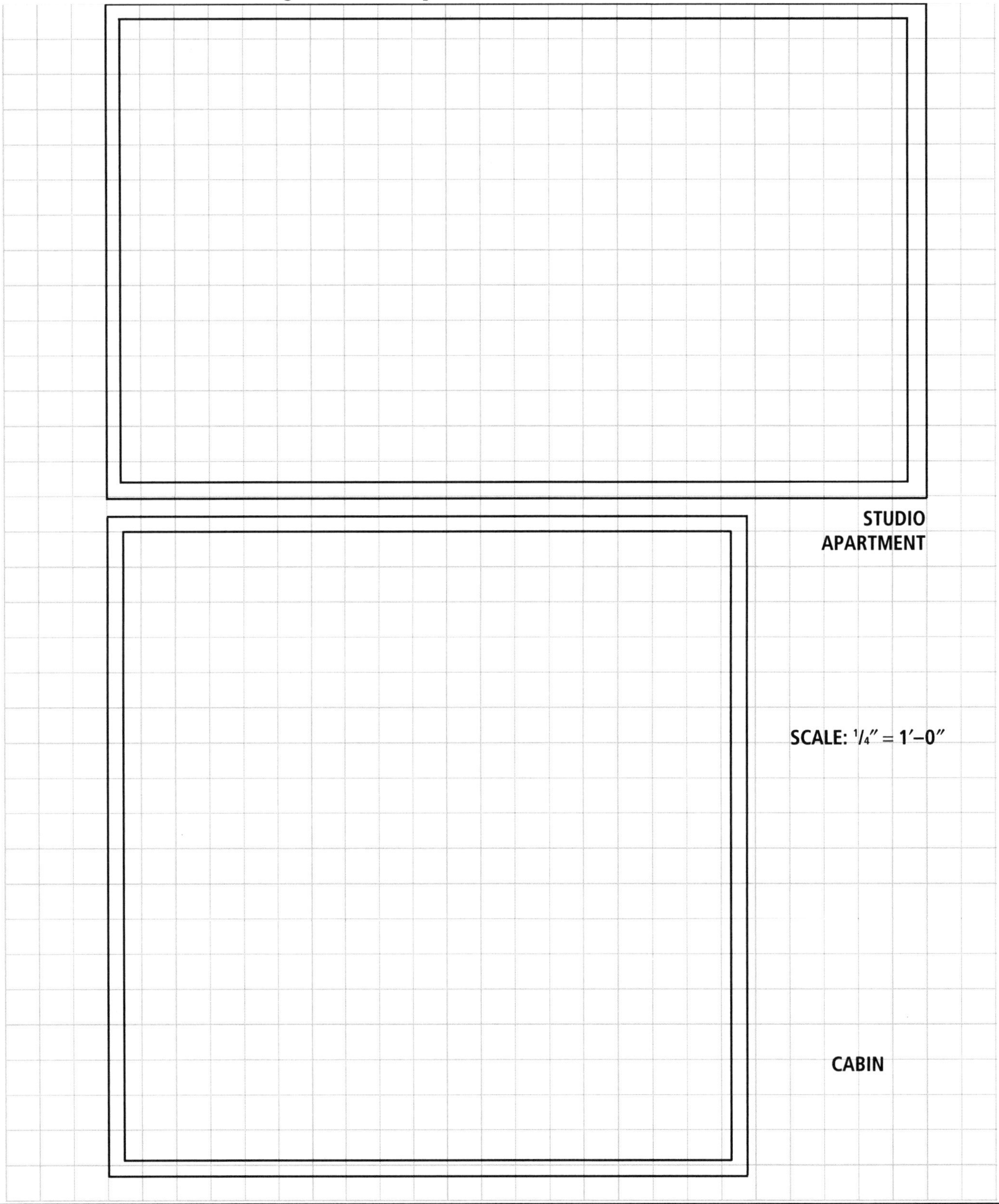

STUDIO APARTMENT

SCALE: ¼″ = 1′–0″

CABIN

TITLE	DRAFTER	SCALE	DATE	DWG NO.

EXERCISE 14-3

Use the grid below to design a bathroom for a person who uses a wheelchair. Follow the design guidelines in Chapter 14 of your text.

| TITLE | DRAFTER | SCALE | DATE | DWG NO. |

EXERCISE 14-4

Use the grid below to design a kitchen for a person who uses a wheelchair. Follow the design guidelines in Chapter 14 of your text.

TITLE	DRAFTER	SCALE	DATE	DWG NO.

EXERCISE 14-5

Using instruments, draw these symbols in the space below.

FLOOR PLAN SYMBOLS

WALL CABINET • FLOOR CABINET • ELECTRIC MOTOR • COOKTOP • RANGE • REFRIGERATOR • DISHWASHER • OVEN

DRYER • WASHER • 2 PC WATER CLOSETS • 1 PC WATER CLOSETS • BIDET • URINAL • COUNTER TOP LAVATORIES • FREE STANDING LAVATORIES • CORNER LAVATORIES

15 R UP — STAIRS (10" TREAD, 7 1/4" RISER) • 6" WALL PLAIN • 6" WALL LINED • 6" WALL SOLID • 8" WALL MASONRY

SCALE: 1/4"=1'-0"

TITLE	DRAFTER	SCALE	DATE	DWG NO.

EXERCISE 14-6

Multiple-choice questions: Select the correct answer and write the letter in the space at the right.

1. On first drafts, it is almost always necessary to	1. _____
   a. improve or revise.	c. leave room for additional plans.
   b. include the dimensions.	d. get copies.

2. The open space in a room where people move about is called the	2. _____
   a. hallway.	c. walkway.
   b. traffic area.	d. entry.

3. The dimensions of a square foot	3. _____
   a. are one foot on all sides.	c. depend on the size of a room.
   b. are one foot by one foot by one foot.	d. are 24 sq. in.

4. The total expenses for creating a building are called the	4. _____
   a. site expenses.	c. building costs.
   b. construction expenses.	d. construction costs.

Copy the room templates. Cut out the templates. Arrange the room templates into a functional plan. Sketch a final floor plan from your arrangement. Add halls, closets, windows, doors, fixtures, and so forth.

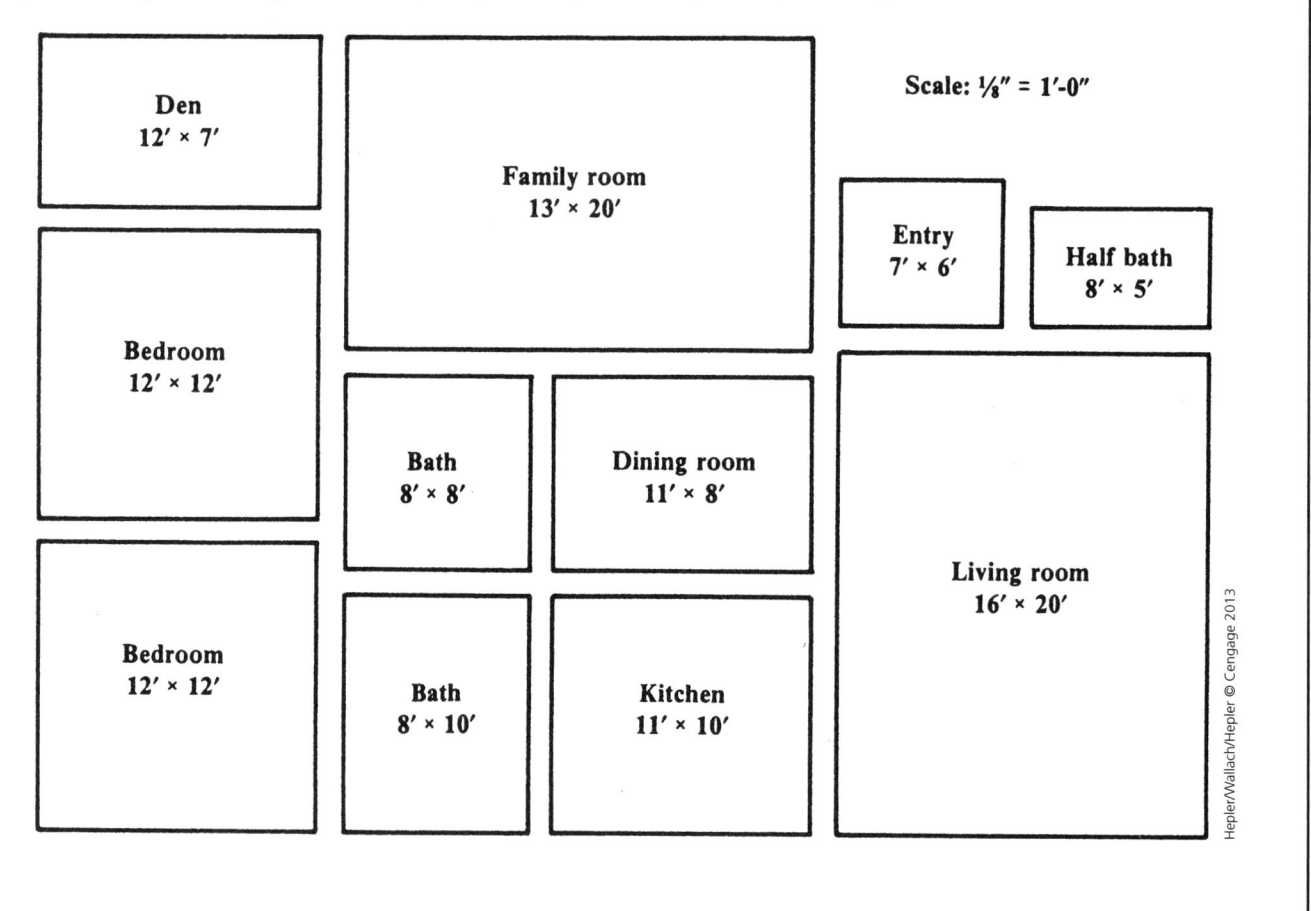

Scale: ⅛" = 1'-0"

Room templates:
- Den 12' × 7'
- Family room 13' × 20'
- Entry 7' × 6'
- Half bath 8' × 5'
- Bedroom 12' × 12'
- Bath 8' × 8'
- Dining room 11' × 8'
- Living room 16' × 20'
- Bedroom 12' × 12'
- Bath 8' × 10'
- Kitchen 11' × 10'

TITLE	DRAFTER	SCALE	DATE	DWG NO.

EXERCISE 14-5

Using instruments, draw these symbols in the space below.

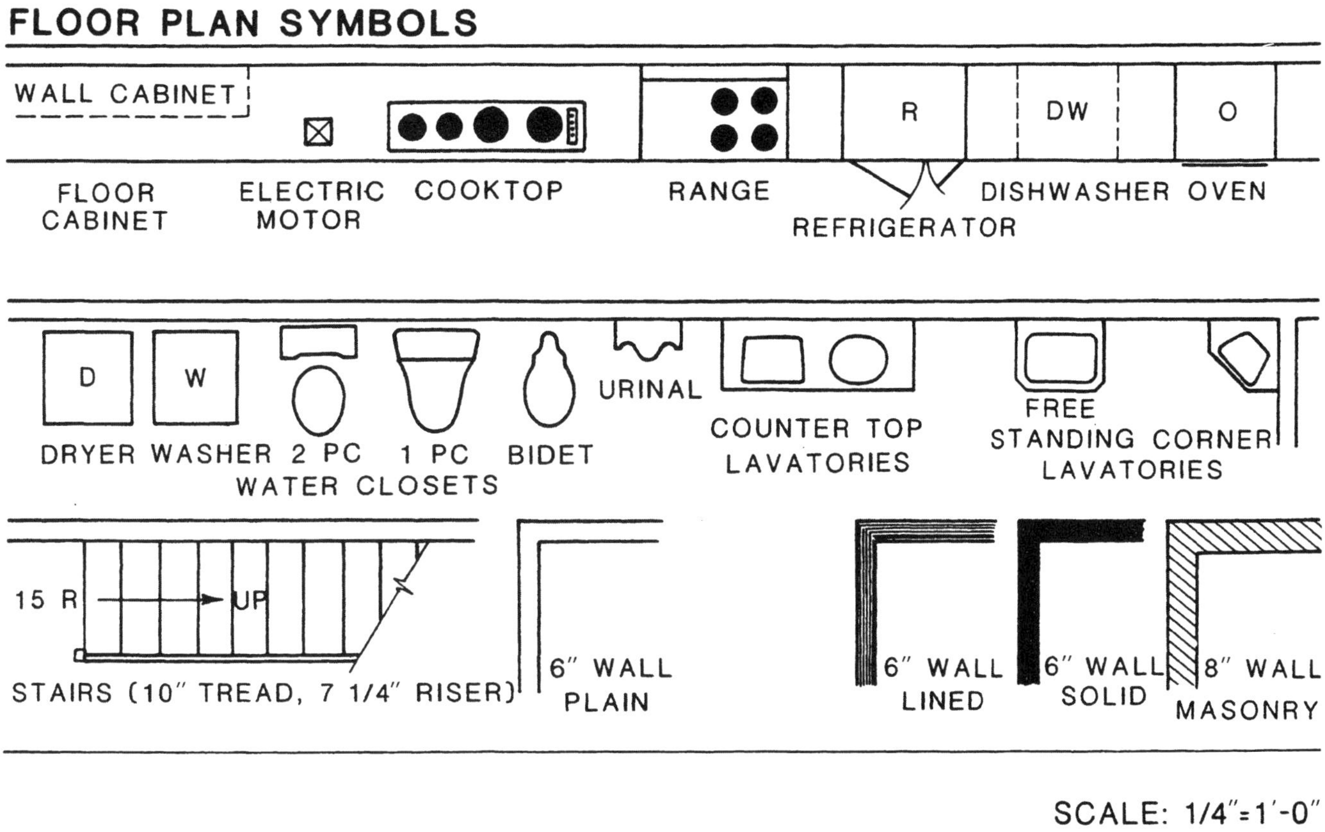

FLOOR PLAN SYMBOLS

WALL CABINET

FLOOR CABINET | ELECTRIC MOTOR | COOKTOP | RANGE | REFRIGERATOR | DISHWASHER | OVEN

DRYER | WASHER | 2 PC / 1 PC WATER CLOSETS | BIDET | URINAL | COUNTER TOP LAVATORIES | FREE STANDING CORNER LAVATORIES

15 R → UP
STAIRS (10" TREAD, 7 1/4" RISER) | 6" WALL PLAIN | 6" WALL LINED | 6" WALL SOLID | 8" WALL MASONRY

SCALE: 1/4"=1'-0"

TITLE	DRAFTER	SCALE	DATE	DWG NO.

EXERCISE 14-6

Multiple-choice questions: Select the correct answer and write the letter in the space at the right.

1. On first drafts, it is almost always necessary to 1. _____
   a. improve or revise.
   b. include the dimensions.
   c. leave room for additional plans.
   d. get copies.

2. The open space in a room where people move about is called the 2. _____
   a. hallway.
   b. traffic area.
   c. walkway.
   d. entry.

3. The dimensions of a square foot 3. _____
   a. are one foot on all sides.
   b. are one foot by one foot by one foot.
   c. depend on the size of a room.
   d. are 24 sq. in.

4. The total expenses for creating a building are called the 4. _____
   a. site expenses.
   b. construction expenses.
   c. building costs.
   d. construction costs.

Copy the room templates. Cut out the templates. Arrange the room templates into a functional plan. Sketch a final floor plan from your arrangement. Add halls, closets, windows, doors, fixtures, and so forth.

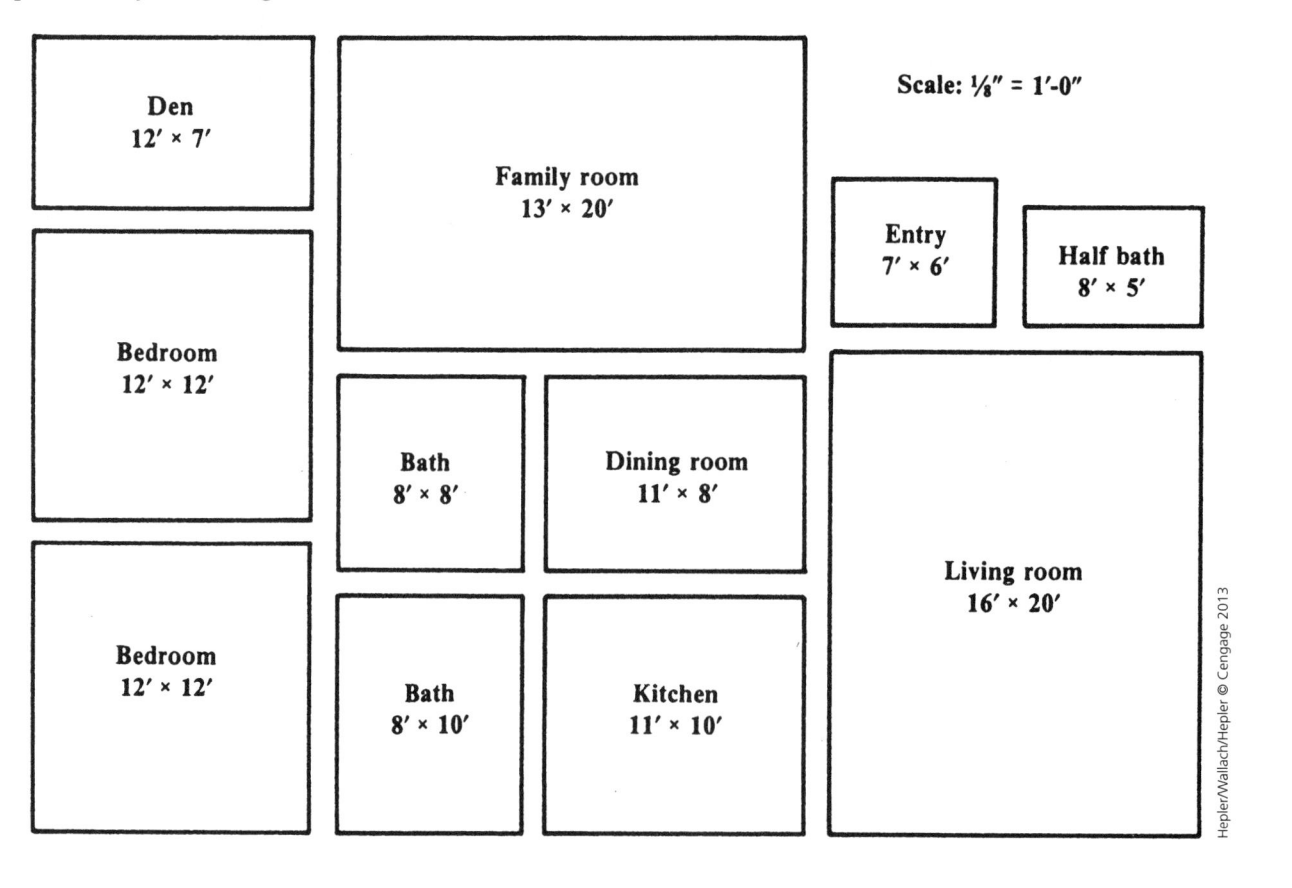

Scale: ⅛" = 1'-0"

Room templates:
- Den 12' × 7'
- Bedroom 12' × 12'
- Bedroom 12' × 12'
- Family room 13' × 20'
- Bath 8' × 8'
- Bath 8' × 10'
- Dining room 11' × 8'
- Kitchen 11' × 10'
- Entry 7' × 6'
- Half bath 8' × 5'
- Living room 16' × 20'

TITLE	DRAFTER	SCALE	DATE	DWG NO.

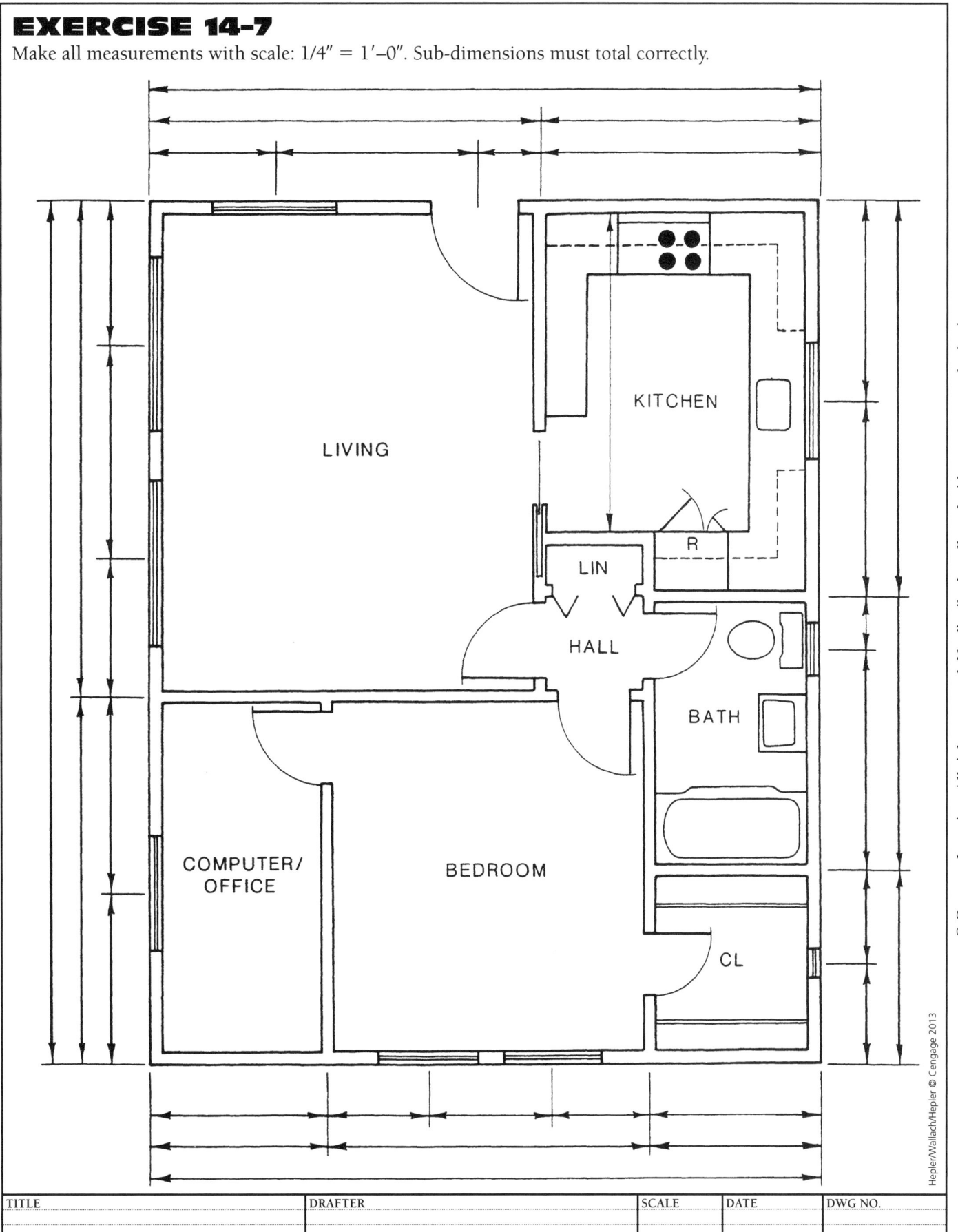

EXERCISE 14-7

Make all measurements with scale: 1/4" = 1'–0". Sub-dimensions must total correctly.

TITLE	DRAFTER	SCALE	DATE	DWG NO.

EXERCISE 15-1

Using the grid below, redraw the floor plan for the first floor of this A-frame house.

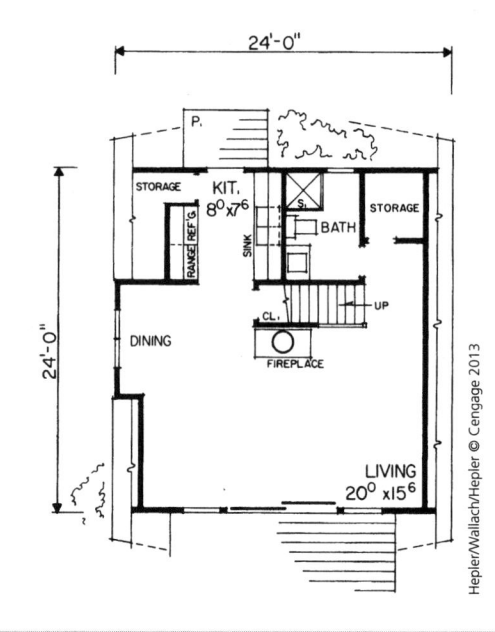

TITLE	DRAFTER	SCALE	DATE	DWG NO.

EXERCISE 15-2

Redraw on the grid below the floor plan for the second floor of the A-frame house.

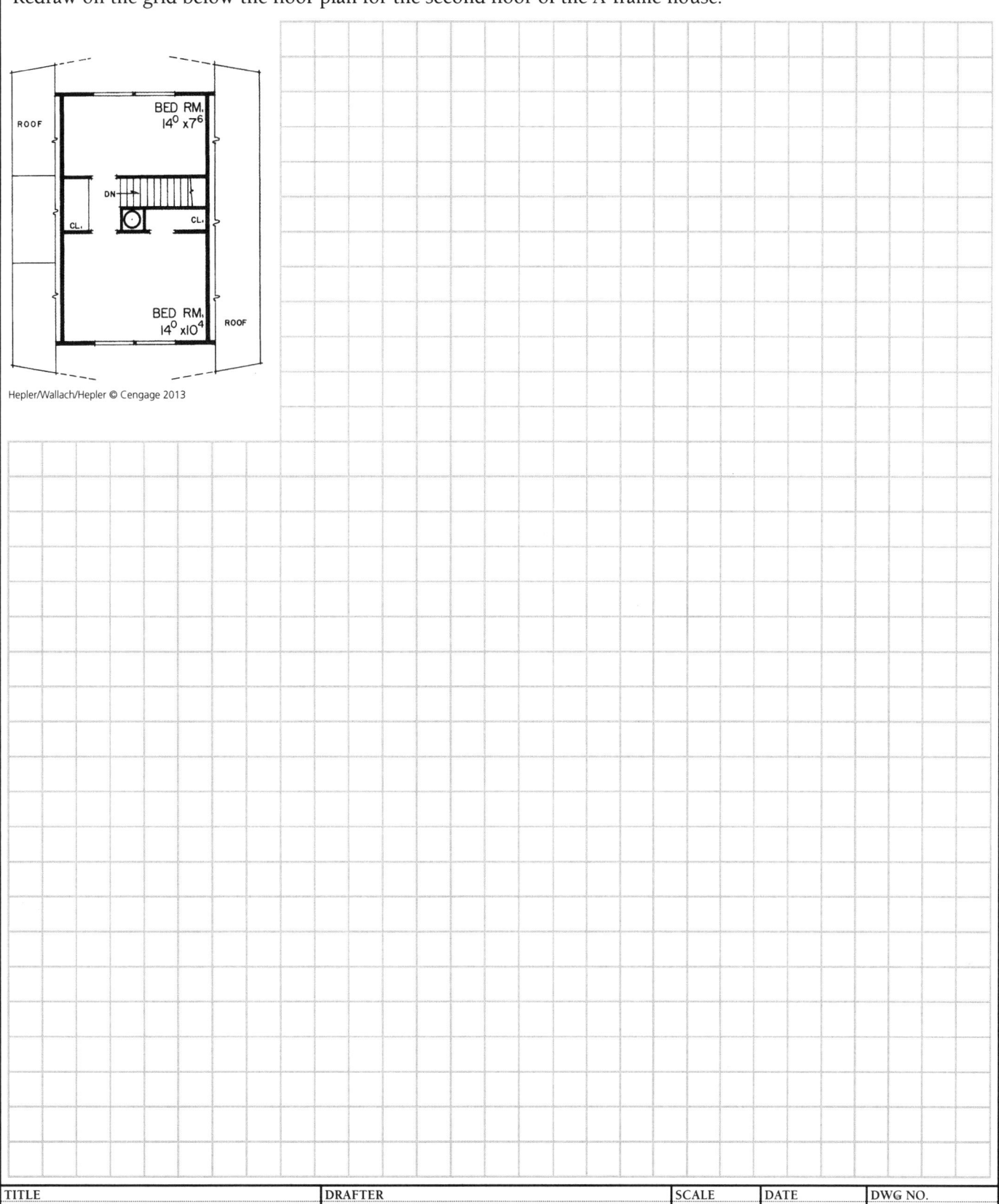

BED RM. $14^0 \times 7^6$

ROOF

DN

CL.   CL.

BED RM. $14^0 \times 10^4$

ROOF

Hepler/Wallach/Hepler © Cengage 2013

TITLE	DRAFTER	SCALE	DATE	DWG NO.

EXERCISE 15-3

Identify the symbols shown in the space at the right.

APPLIANCE SYMBOLS

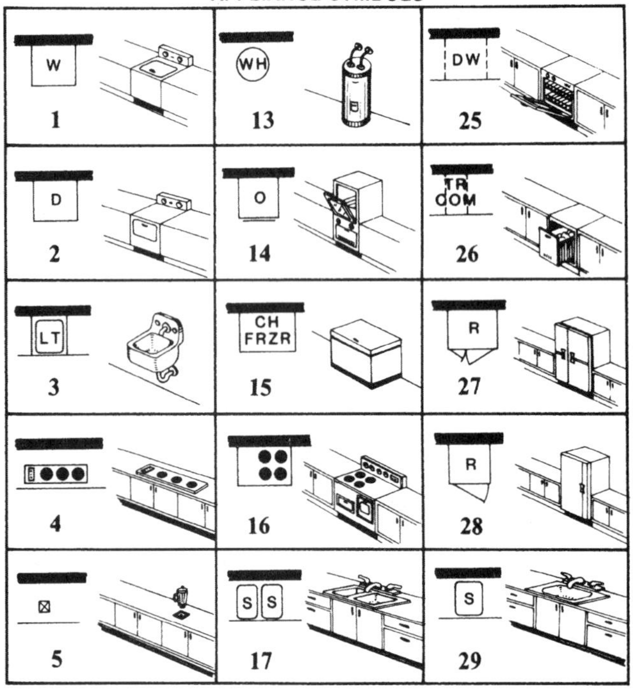

PLUMBING SYMBOLS

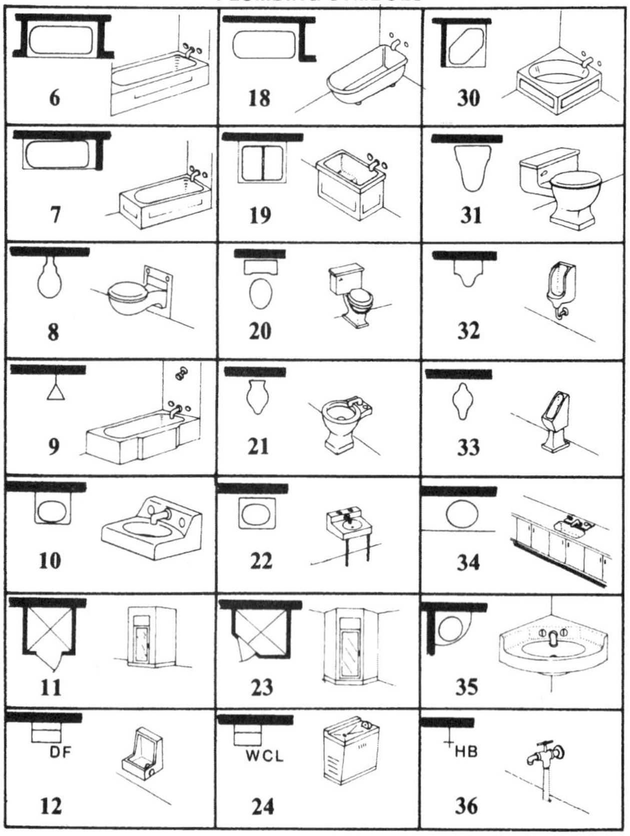

1. _____
2. _____
3. _____
4. _____
5. _____
6. _____
7. _____
8. _____
9. _____
10. _____
11. _____
12. _____
13. _____
14. _____
15. _____
16. _____
17. _____
18. _____
19. _____
20. _____
21. _____
22. _____
23. _____
24. _____
25. _____
26. _____
27. _____
28. _____
29. _____
30. _____
31. _____
32. _____
33. _____
34. _____
35. _____
36. _____

TITLE	DRAFTER	SCALE	DATE	DWG NO.

EXERCISE 15-4

Multiple-choice questions: Select the correct answer and write the letter in the space at the right.

1. Something used to represent another item is called a(n)   1. _____
   a. anagram.
   b. synonym.
   c. acronym.
   d. symbol.

2. A drawing that shows vertical sections, such as walls of a building, is called a   2. _____
   a. vertical drawing.
   b. elevation drawing.
   c. horizontal drawing.
   d. height drawing.

3. Certain letters of architectural terms that represent words are called   3. _____
   a. abbreviations.
   b. synonyms.
   c. architectural symbols.
   d. acronyms.

Name each architectural symbol:

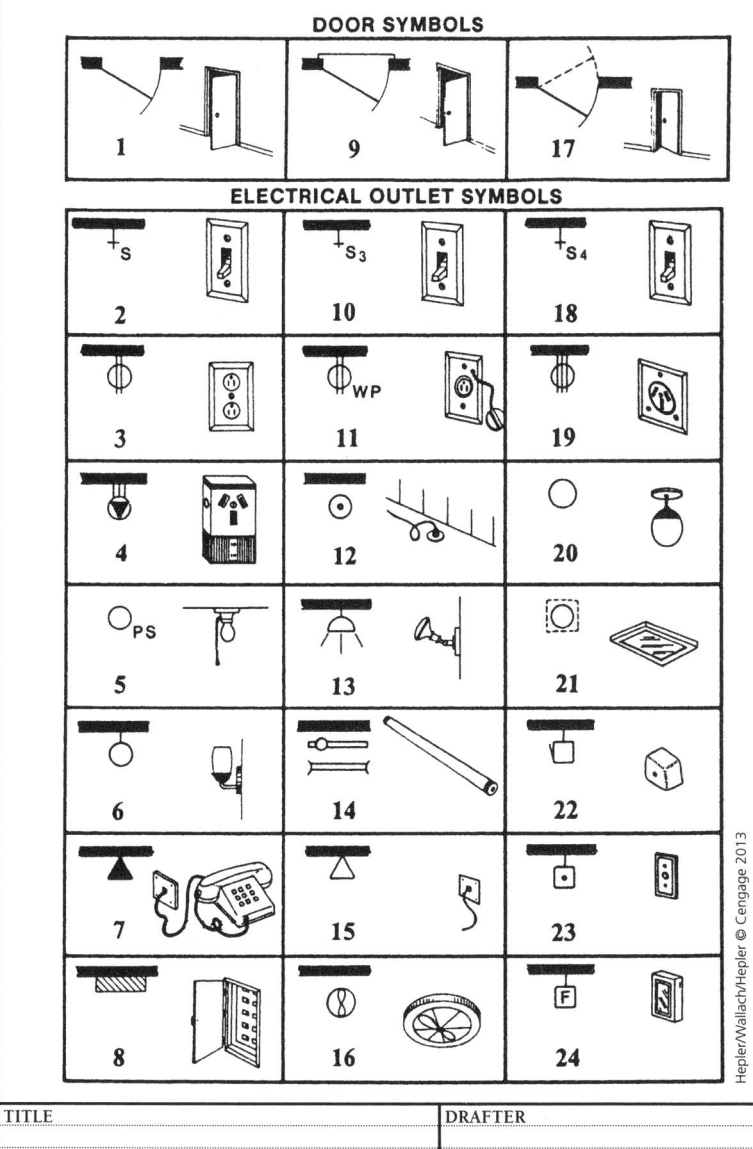

1. _____
2. _____
3. _____
4. _____
5. _____
6. _____
7. _____
8. _____
9. _____
10. _____
11. _____
12. _____
13. _____
14. _____
15. _____
16. _____
17. _____
18. _____
19. _____
20. _____
21. _____
22. _____
23. _____
24. _____

TITLE	DRAFTER	SCALE	DATE	DWG NO.

EXERCISE 15-5

Add fixtures, appliances, windows, doors, and furniture. Figure the square footage and building costs at $150 per square foot. Scale: 1/4″ = 1′–0″.

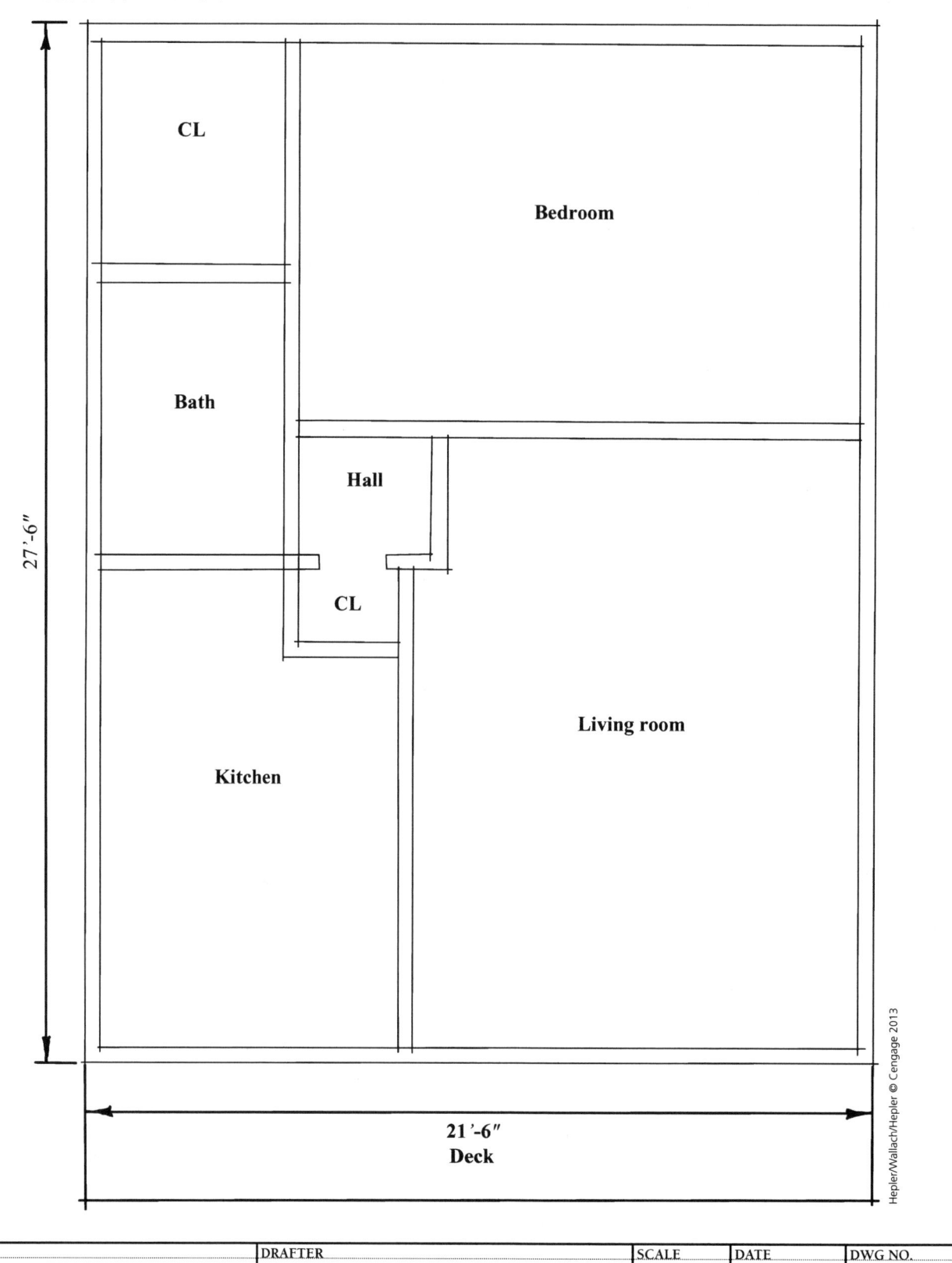

CL

Bedroom

Bath

Hall

CL

Living room

Kitchen

27′-6″

21′-6″
Deck

TITLE	DRAFTER	SCALE	DATE	DWG NO.

EXERCISE 16-5

Refer to the plan below and draw the four elevations with the cabinets and appliances shown.

INTERIOR ELEVATION EXERCISES

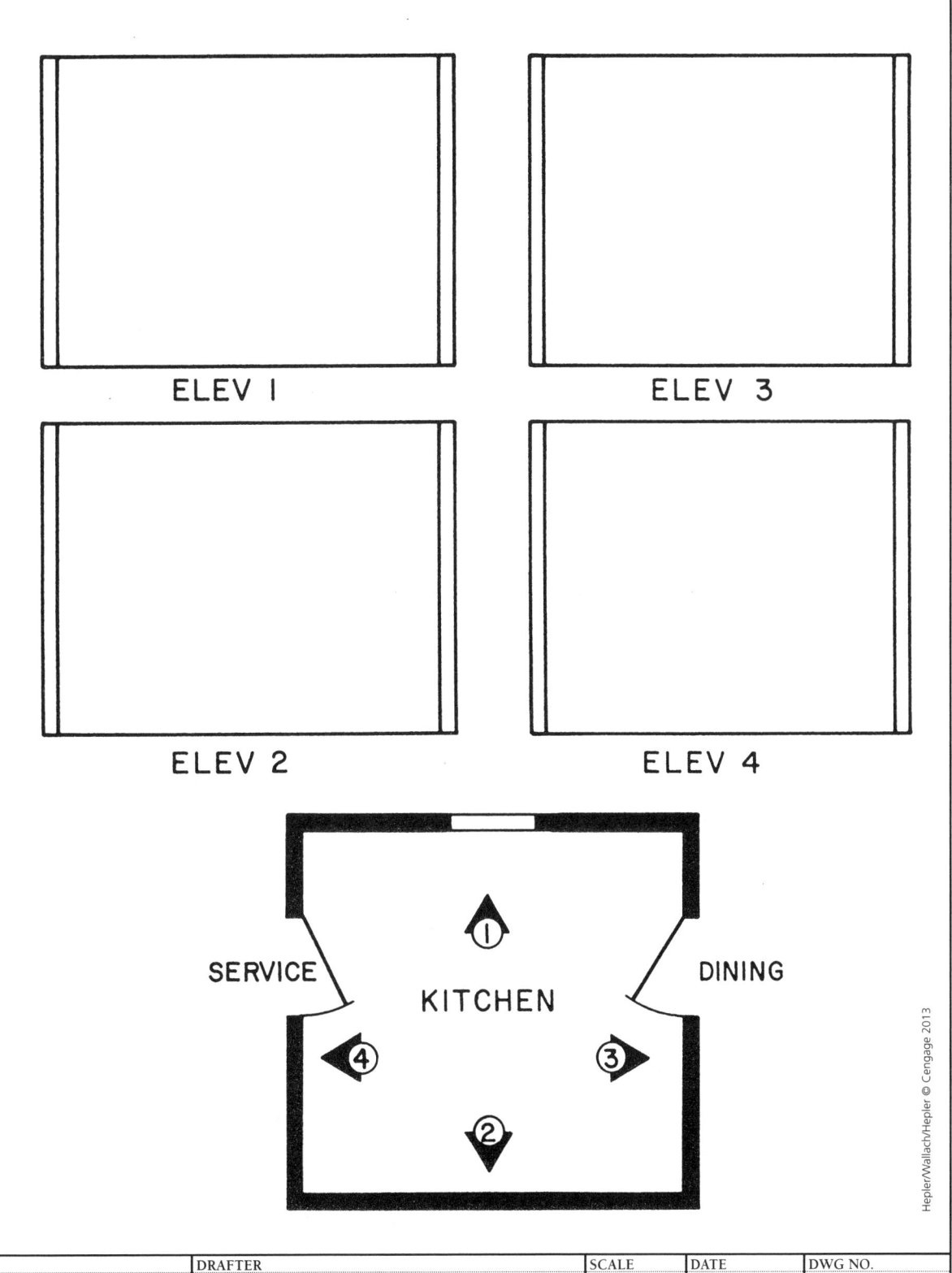

ELEV 1

ELEV 3

ELEV 2

ELEV 4

SERVICE   KITCHEN   DINING

| TITLE | DRAFTER | SCALE | DATE | DWG NO. |

EXERCISE 16-6

Refer to the plan below and draw the four elevations with the furniture shown.

INTERIOR ELEVATION EXERCISES

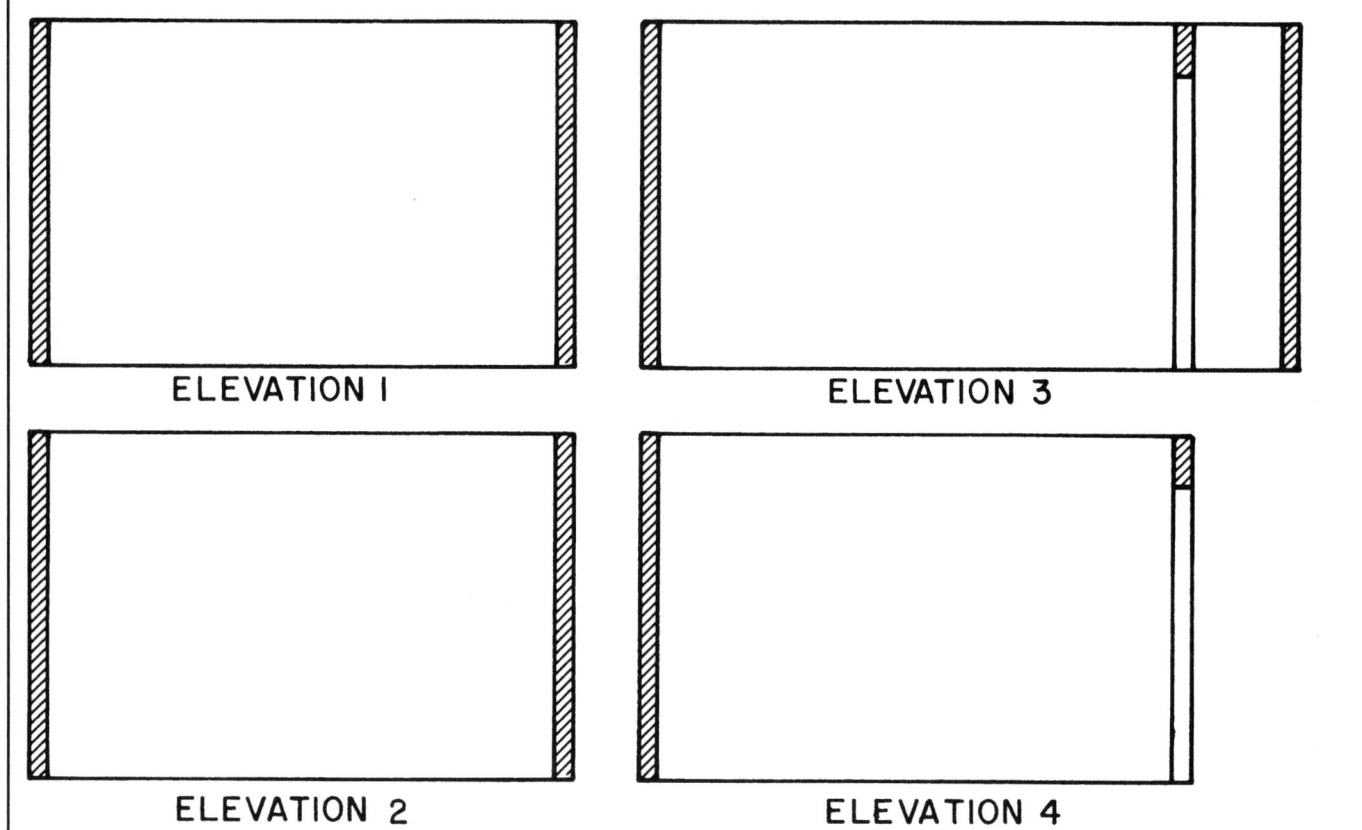

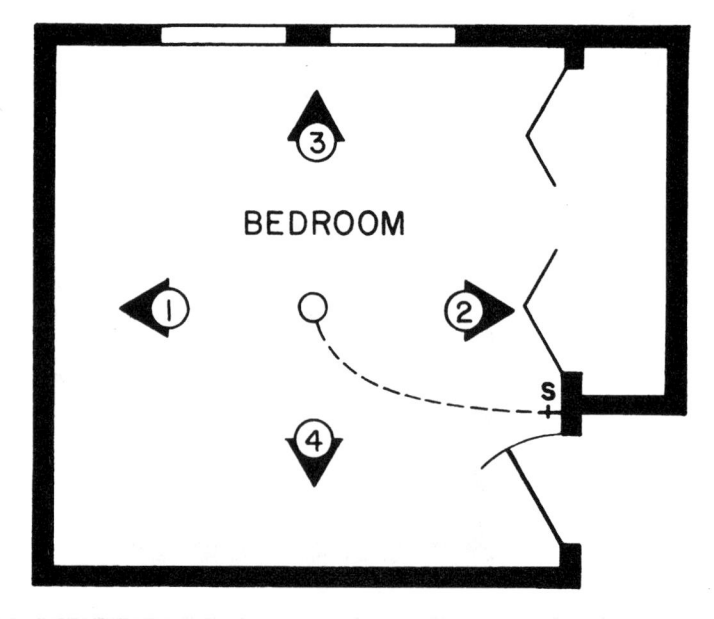

ELEVATION 1

ELEVATION 3

ELEVATION 2

ELEVATION 4

BEDROOM

| TITLE | DRAFTER | SCALE | DATE | DWG NO. |

EXERCISE 16-7

Refer to the plan below and draw the four elevations with the furniture shown.

INTERIOR ELEVATION EXERCISES

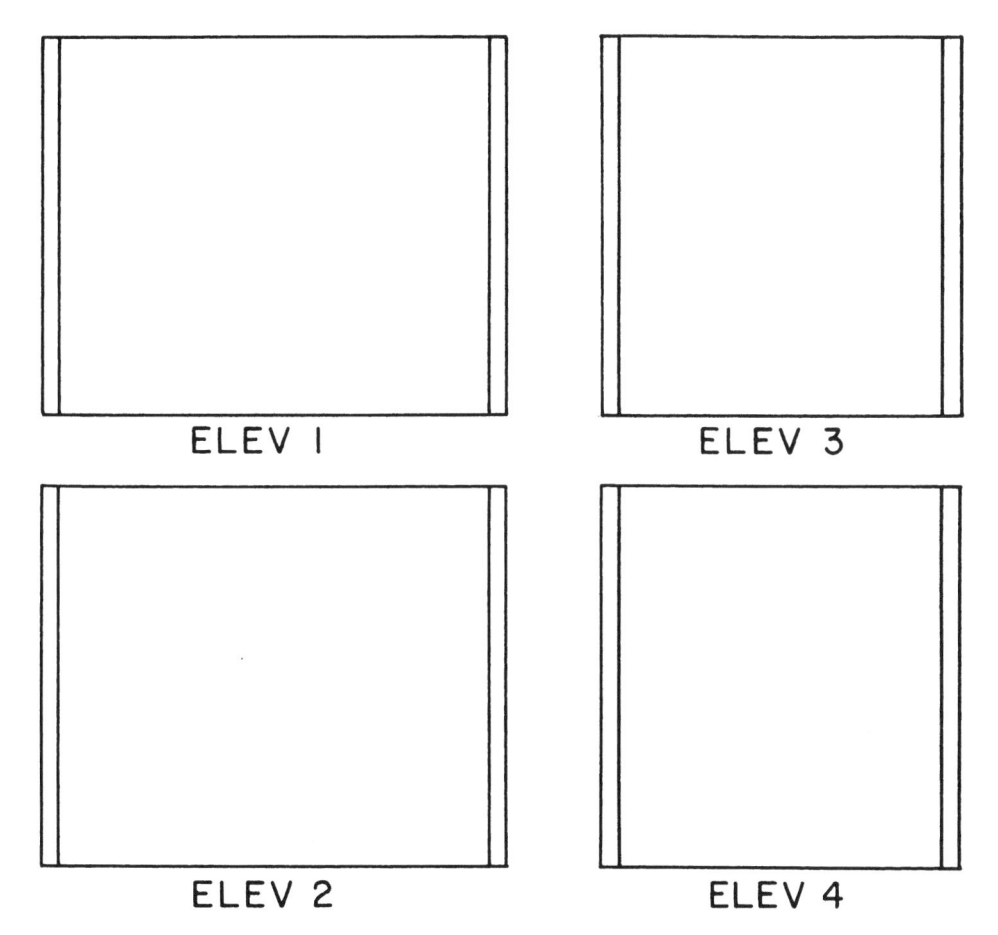

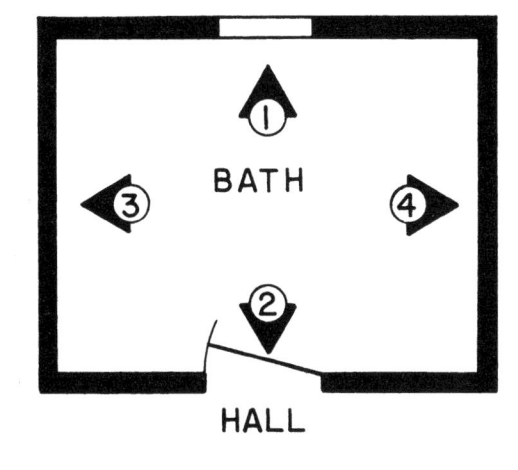

ELEV 1

ELEV 3

ELEV 2

ELEV 4

BATH

HALL

| TITLE | DRAFTER | SCALE | DATE | DWG NO. |

EXERCISE 17-1

Using a scale of 1/4″ = 1′–0″, design and draw a full front elevation for this floor plan.

TITLE	DRAFTER	SCALE	DATE	DWG NO.

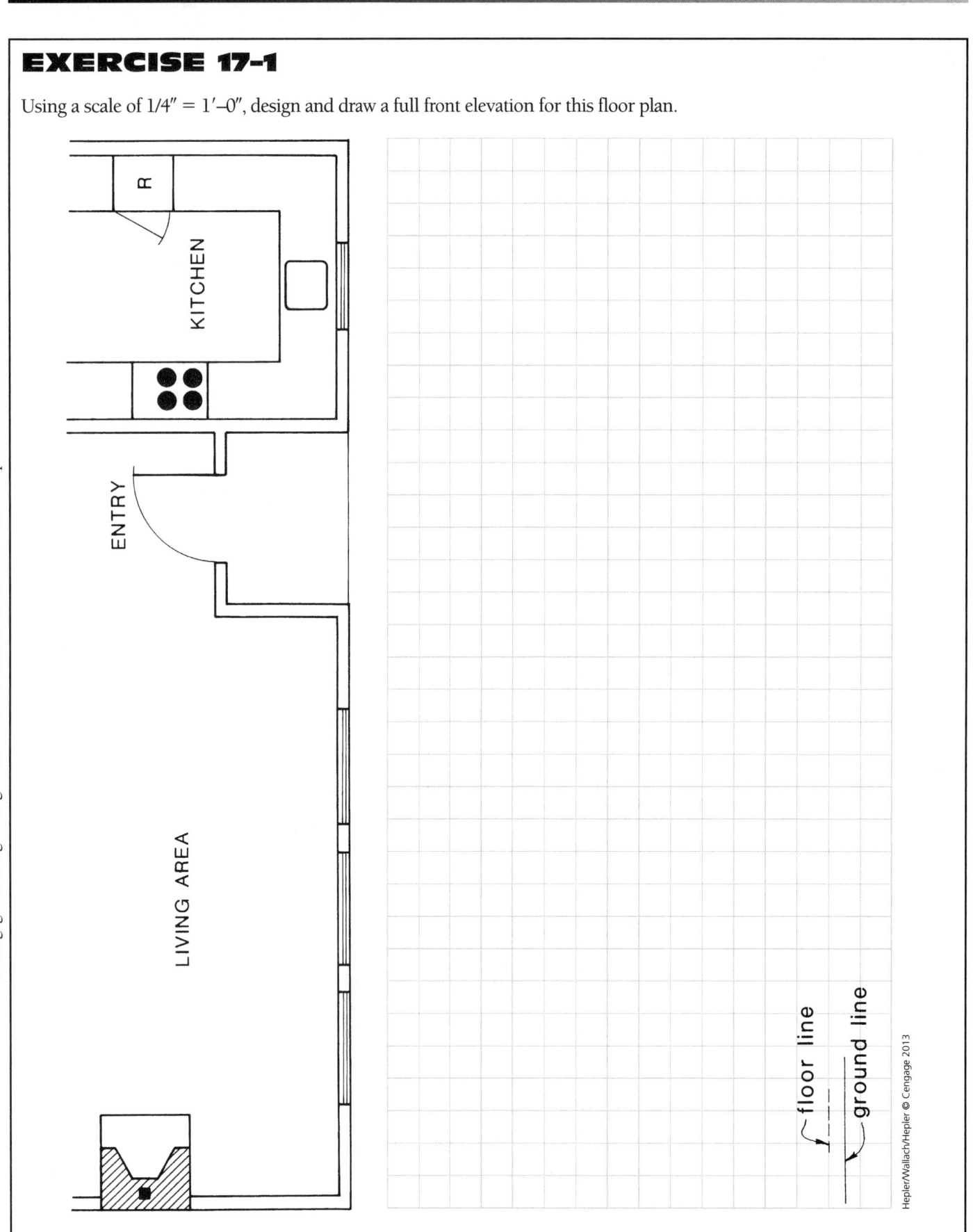

EXERCISE 17 Drawing Elevations ■ **79**

EXERCISE 17-2 Design and draw front elevations in the different styles indicated. All vertical elevation lines should align with the floor plan at the top. Scale: 1/8" = 1'–0".

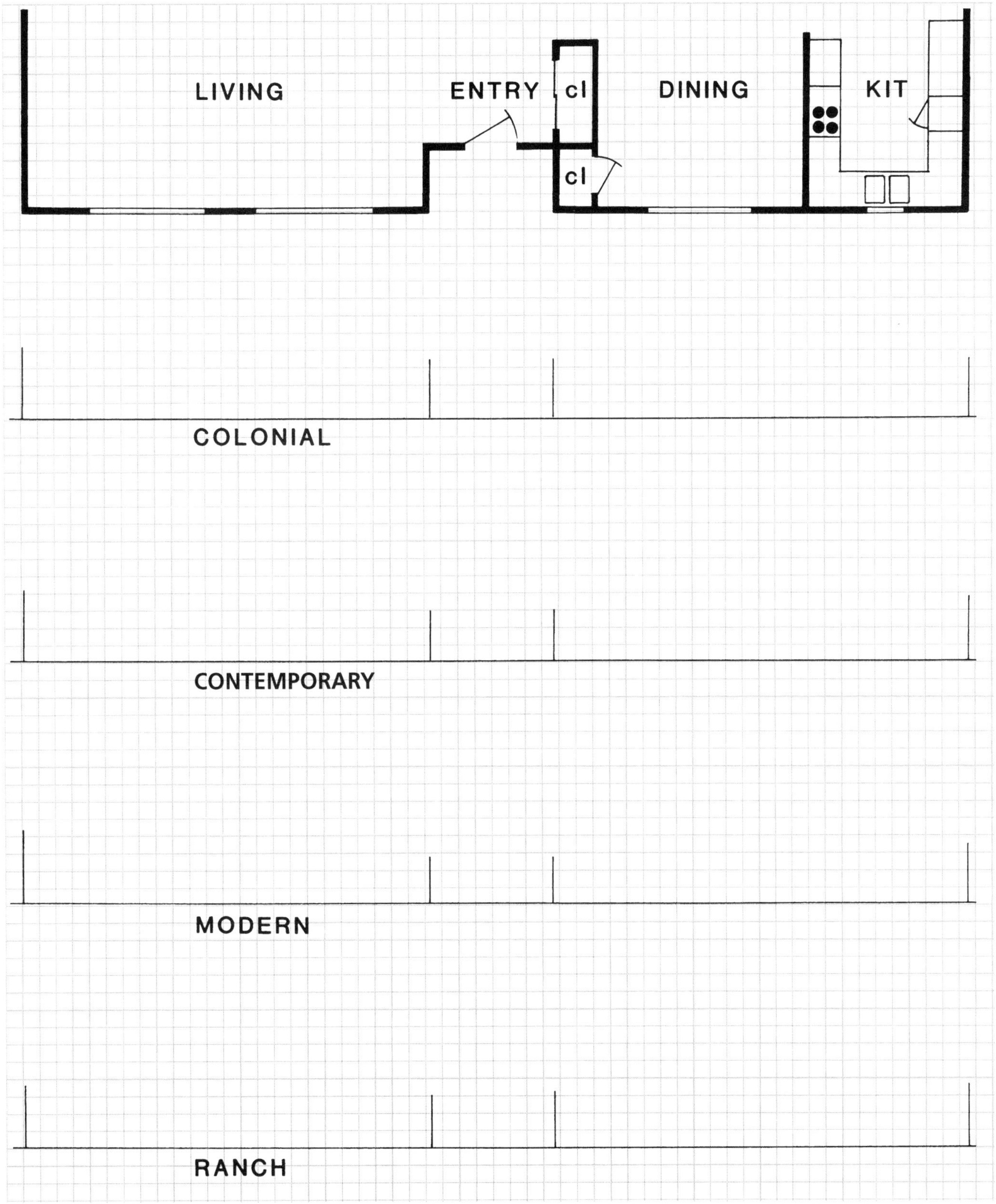

COLONIAL

CONTEMPORARY

MODERN

RANCH

TITLE	DRAFTER	SCALE	DATE	DWG NO.

EXERCISE 17-3

Below is an outline of a cabin floor plan. Complete the floor plan. Then project the four elevations below.
Scale: 1/8″ = 1′–0″.

FLOOR PLAN 36′-0″x20′-0″

NORTH ELEVATION   **EAST ELEVATION**

SOUTH ELEVATION   **WEST ELEVATION**

| TITLE | DRAFTER | SCALE | DATE | DWG NO. |

EXERCISE 17-4

Using a different architectural style for each, complete the elevations. Show windows, doors, roof styles, materials, and chimneys as needed. Scale: 1/8 = 1'–0".

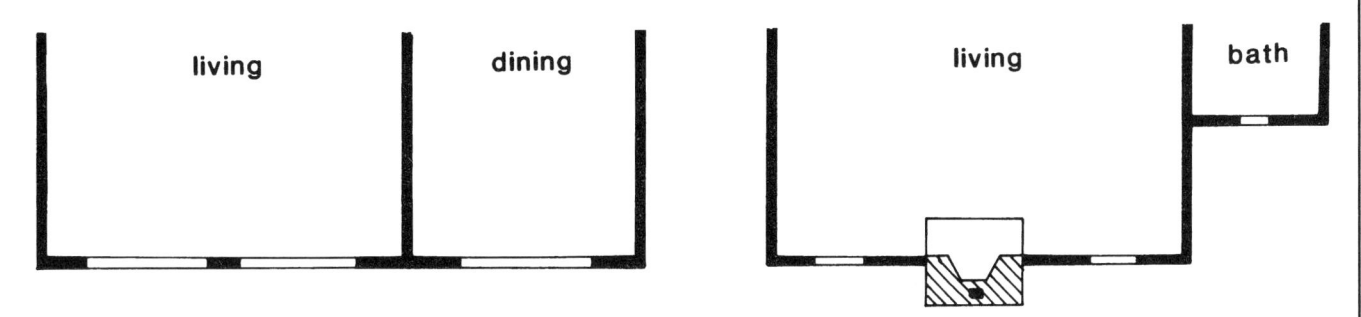

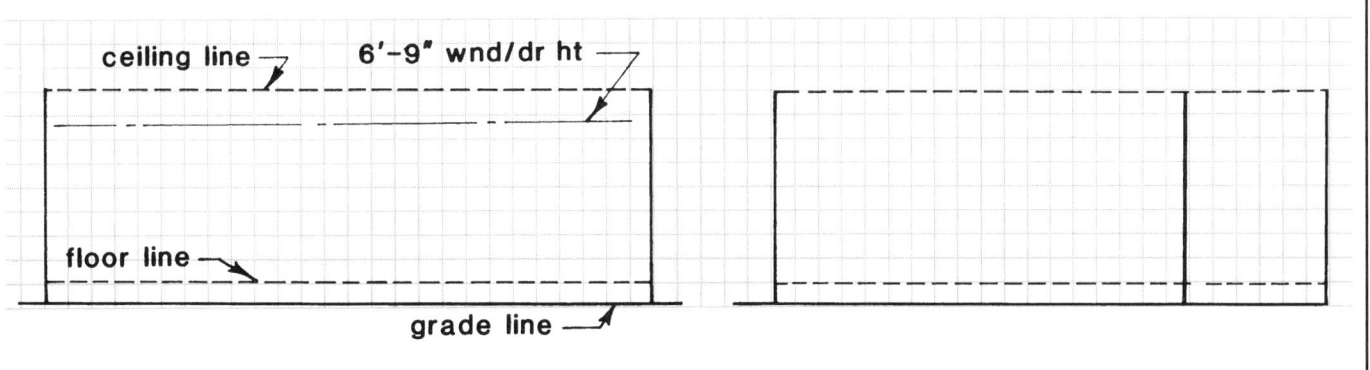

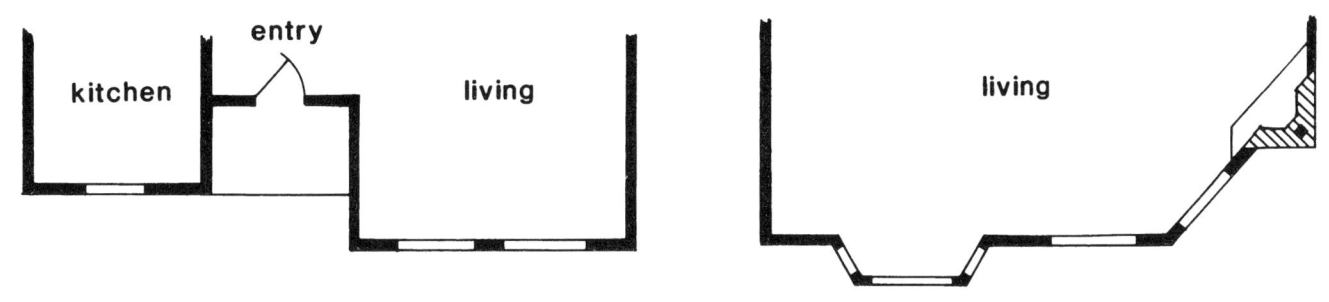

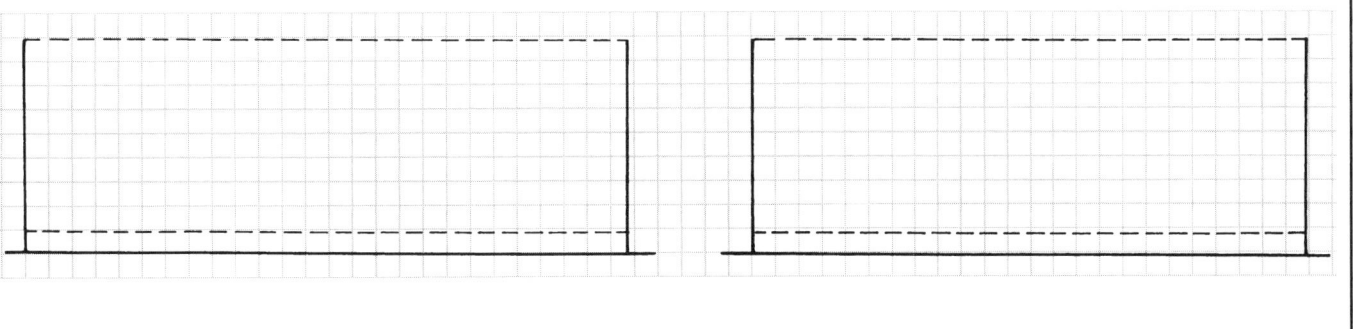

| TITLE | DRAFTER | SCALE | DATE | DWG NO. |

EXERCISE 17-5

Write the names of the window types in the blanks.

PLAN SYMBOL	ELEVATION SYMBOL	PICTORIAL	
			1. _____
			2. _____
			3. _____
			4. _____
			5. _____
			6. _____
			7. _____

TITLE	DRAFTER	SCALE	DATE	DWG NO.

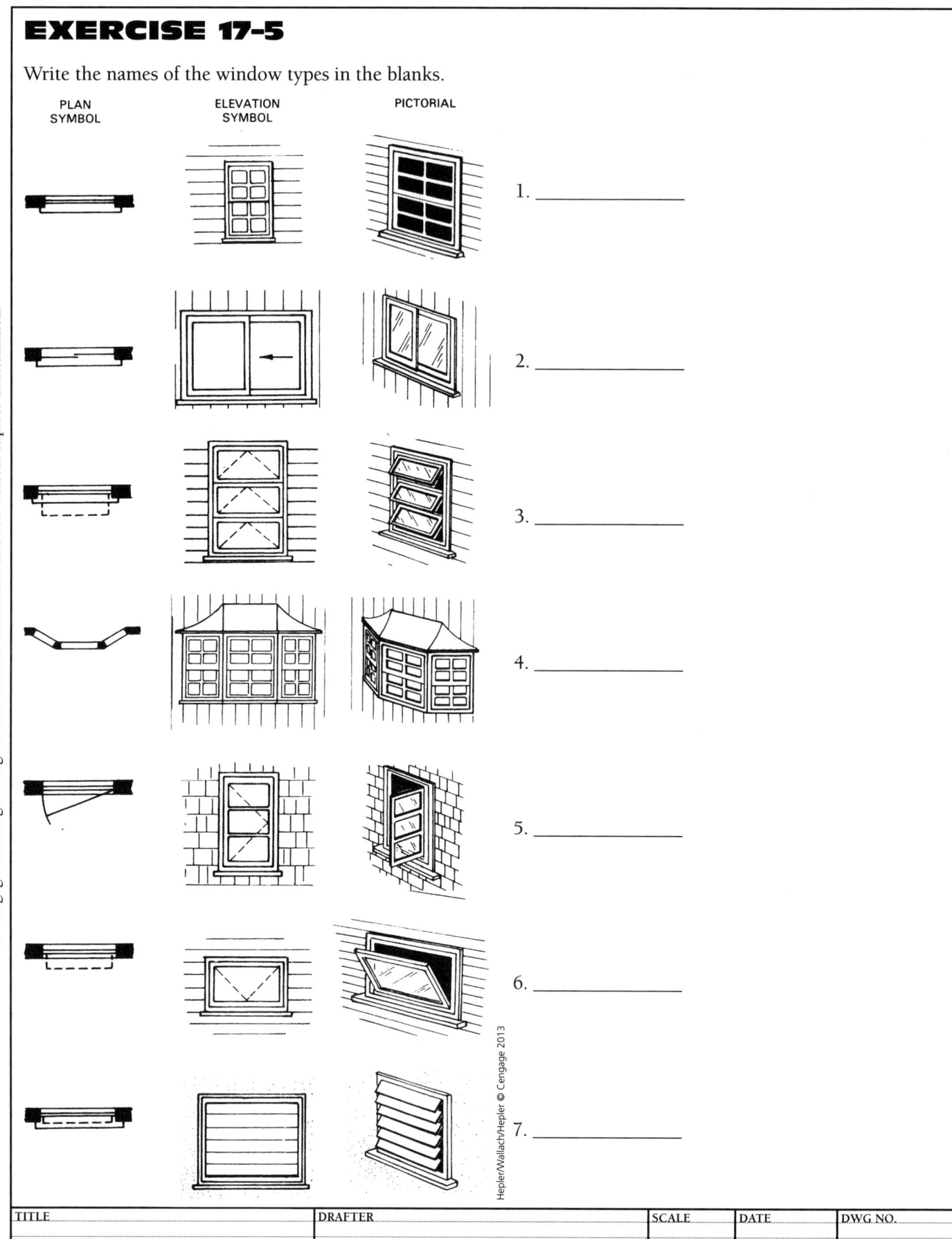

EXERCISE 17-6

Write the letter of the symbols shown in the left-hand column next to the correct wall material in the blank provided.

SYMBOL

SECTION | ELEVATION

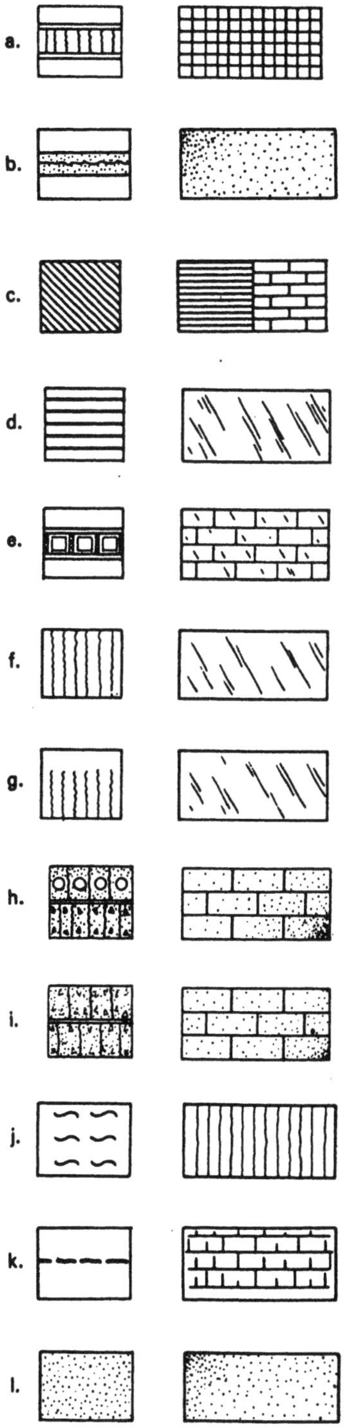

a.
b.
c.
d.
e.
f.
g.
h.
i.
j.
k.
l.

WALL MATERIAL

1. Cast Block _____
2. Composition Shingle _____
3. Concrete Block _____
4. Fabric _____
5. Face Brick _____
6. Frosted Glass _____
7. Glass _____
8. Glass Block _____
9. Glazed Face-Hollow Tile _____
10. Plaster Wall _____
11. Plaster Wall and Metal Lathe _____
12. Structural Glass _____

TITLE	DRAFTER	SCALE	DATE	DWG NO.

EXERCISE 17-7

Refer to the drawings at the top and draw the plan and elevations of either the gambrel or mansard roof building.

GAMBREL ROOF

MANSARD ROOF

FRONT ELEVATION

SIDE ELEVATION

ROOF PLAN

FLOOR PLAN OUTLINE
30'-0" x 20'-0"

TITLE	DRAFTER	SCALE	DATE	DWG NO.

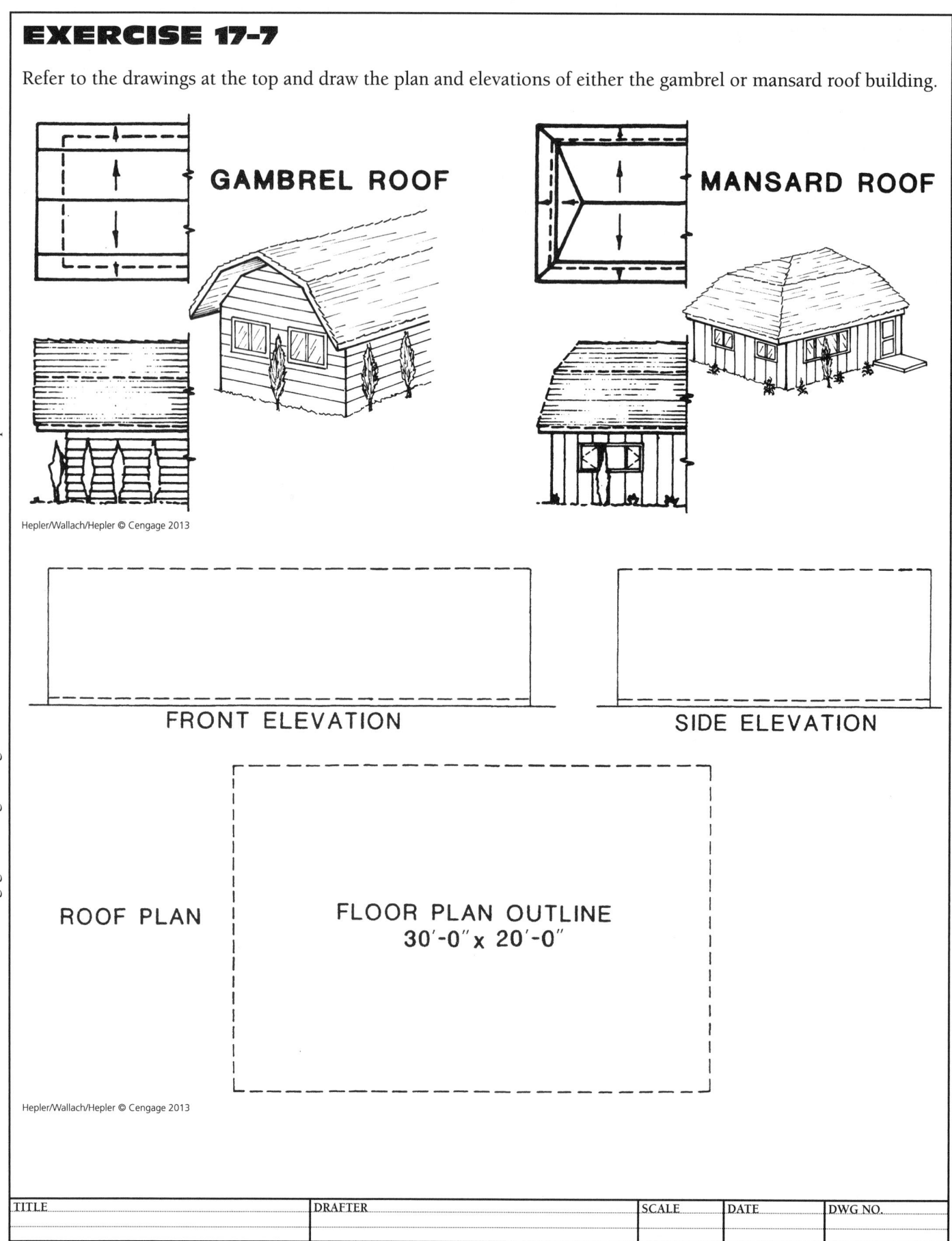

EXERCISE 17-8

Refer to the drawings at the top and draw the plan and elevations of a hip or Dutch hip roof building.

HIP ROOF

DUTCH HIP ROOF

FRONT ELEVATION

SIDE ELEVATION

ROOF PLAN

FLOOR PLAN OUTLINE
30'-0" x 20'-0"

TITLE	DRAFTER	SCALE	DATE	DWG NO.

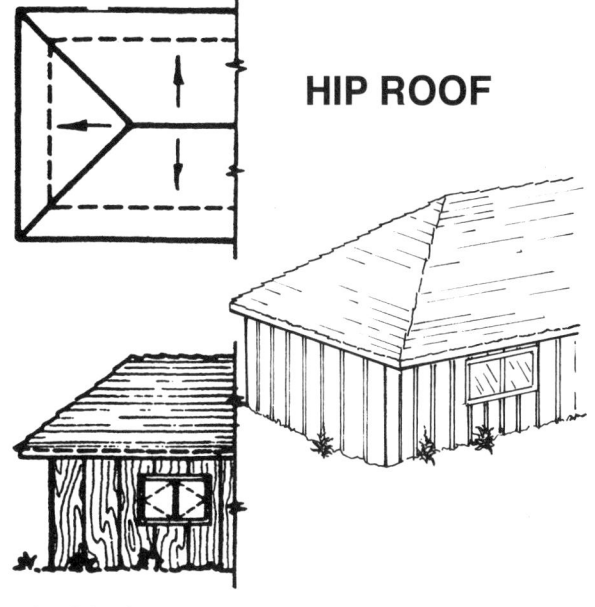

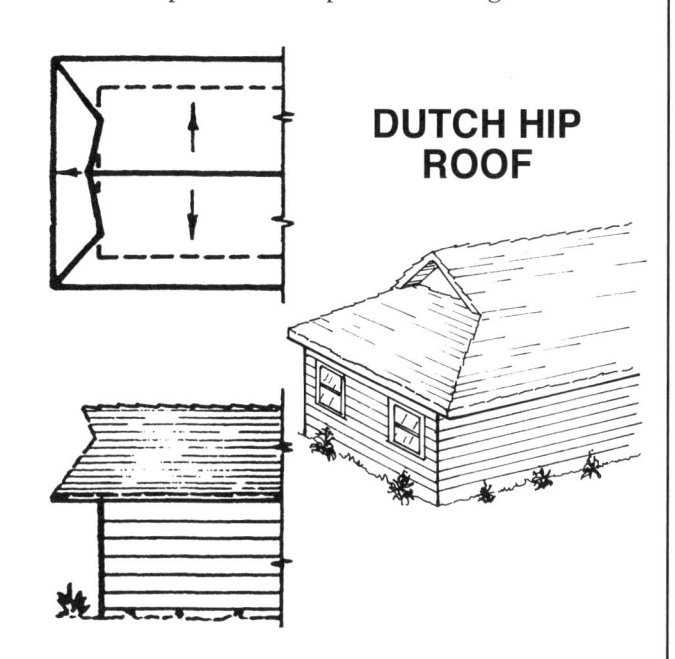

EXERCISE 17-9

Refer to the top drawings and draw the plan and elevations of the gable roof with valley or the gable roof with dormer.

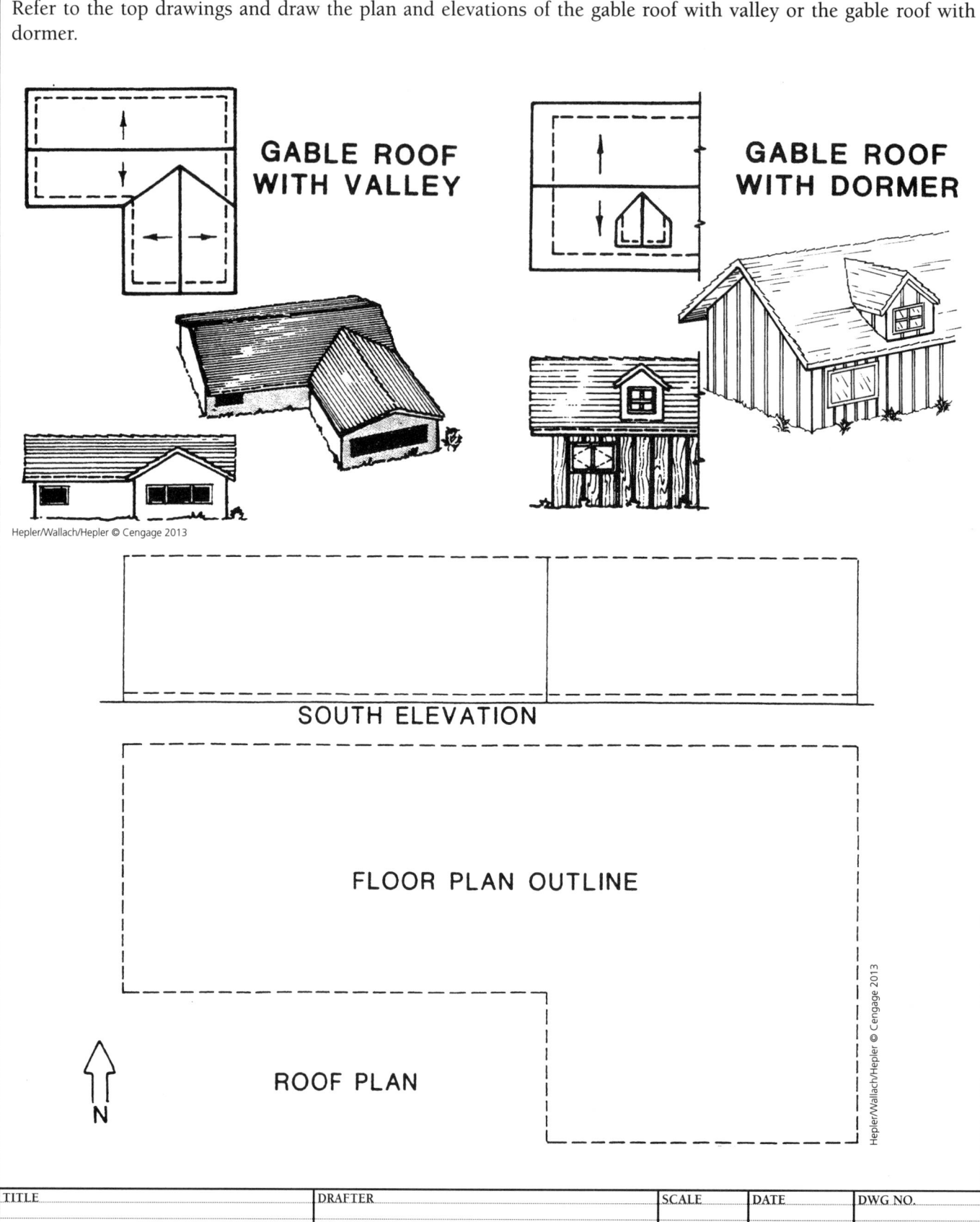

GABLE ROOF WITH VALLEY

GABLE ROOF WITH DORMER

Hepler/Wallach/Hepler © Cengage 2013

SOUTH ELEVATION

FLOOR PLAN OUTLINE

N

ROOF PLAN

Hepler/Wallach/Hepler © Cengage 2013

TITLE	DRAFTER	SCALE	DATE	DWG NO.

EXERCISE 17-10

Refer to the top drawings and draw the plan and elevations of a gable or shed roof building.

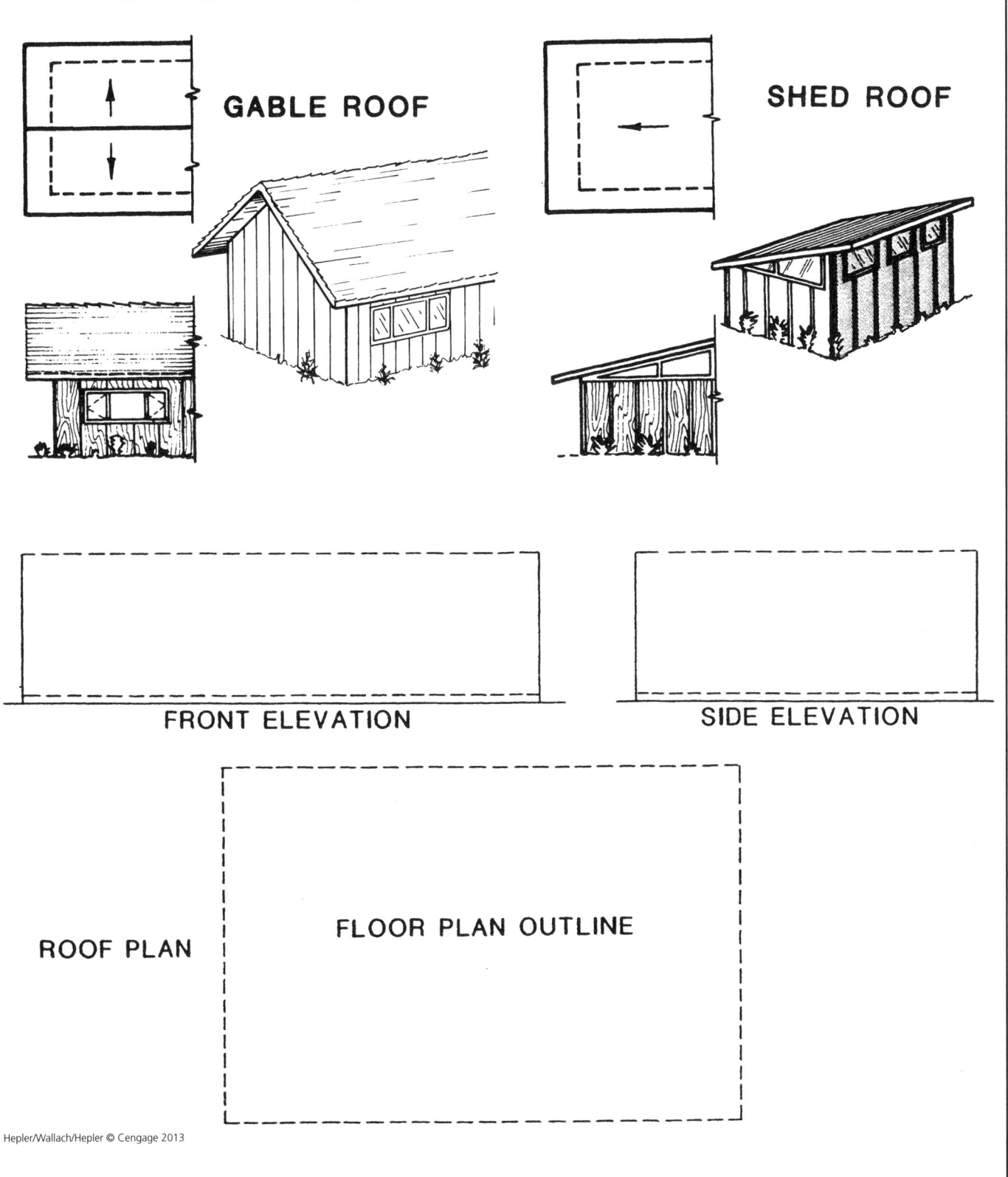

GABLE ROOF

SHED ROOF

FRONT ELEVATION

SIDE ELEVATION

ROOF PLAN

FLOOR PLAN OUTLINE

TITLE	DRAFTER	SCALE	DATE	DWG NO.

# EXERCISE 17-11

Refer to the top drawings and complete the plans and elevations below.

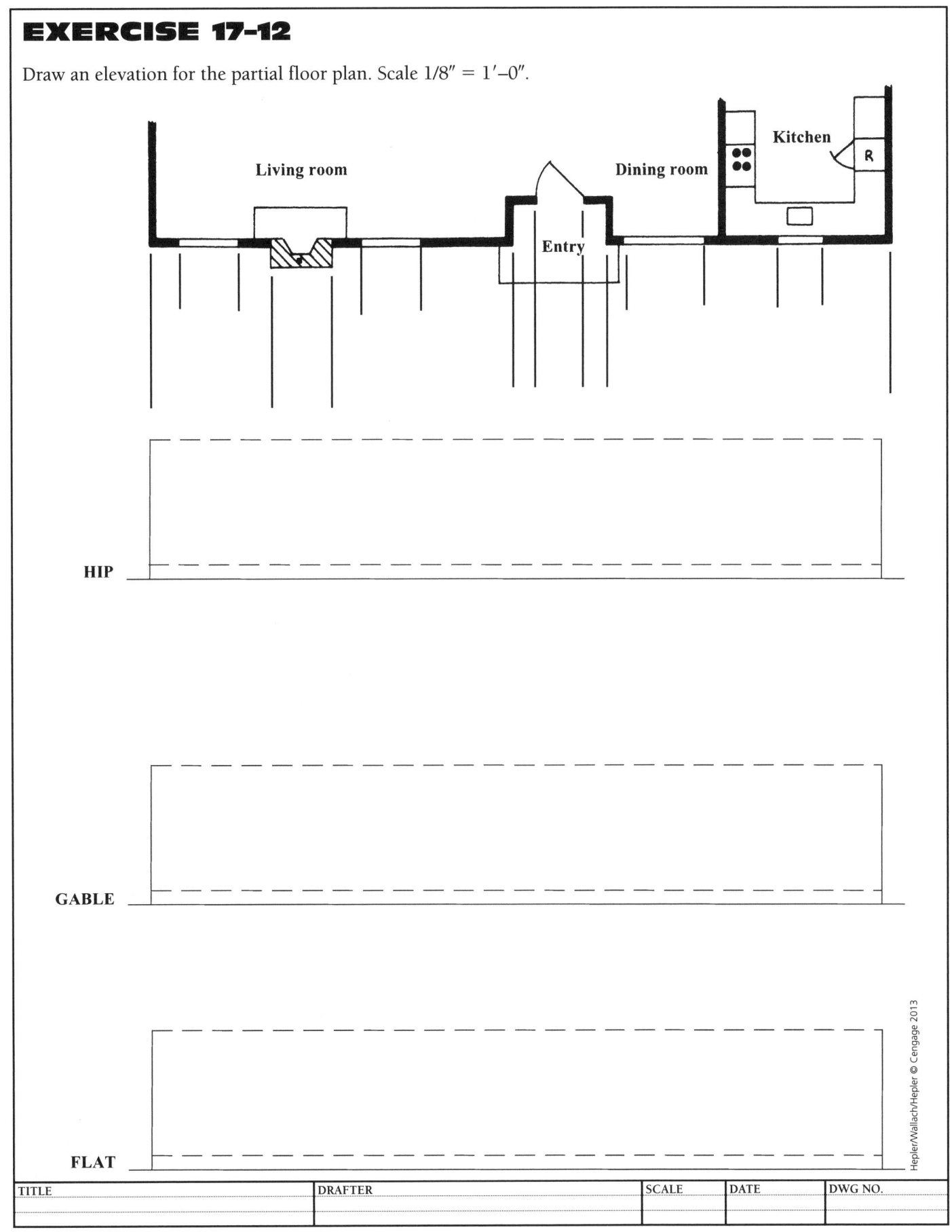

EXERCISE 17 Drawing Elevations ■ **89**

EXERCISE 17-12

Draw an elevation for the partial floor plan. Scale 1/8" = 1'–0".

HIP

GABLE

FLAT

TITLE	DRAFTER	SCALE	DATE	DWG NO.

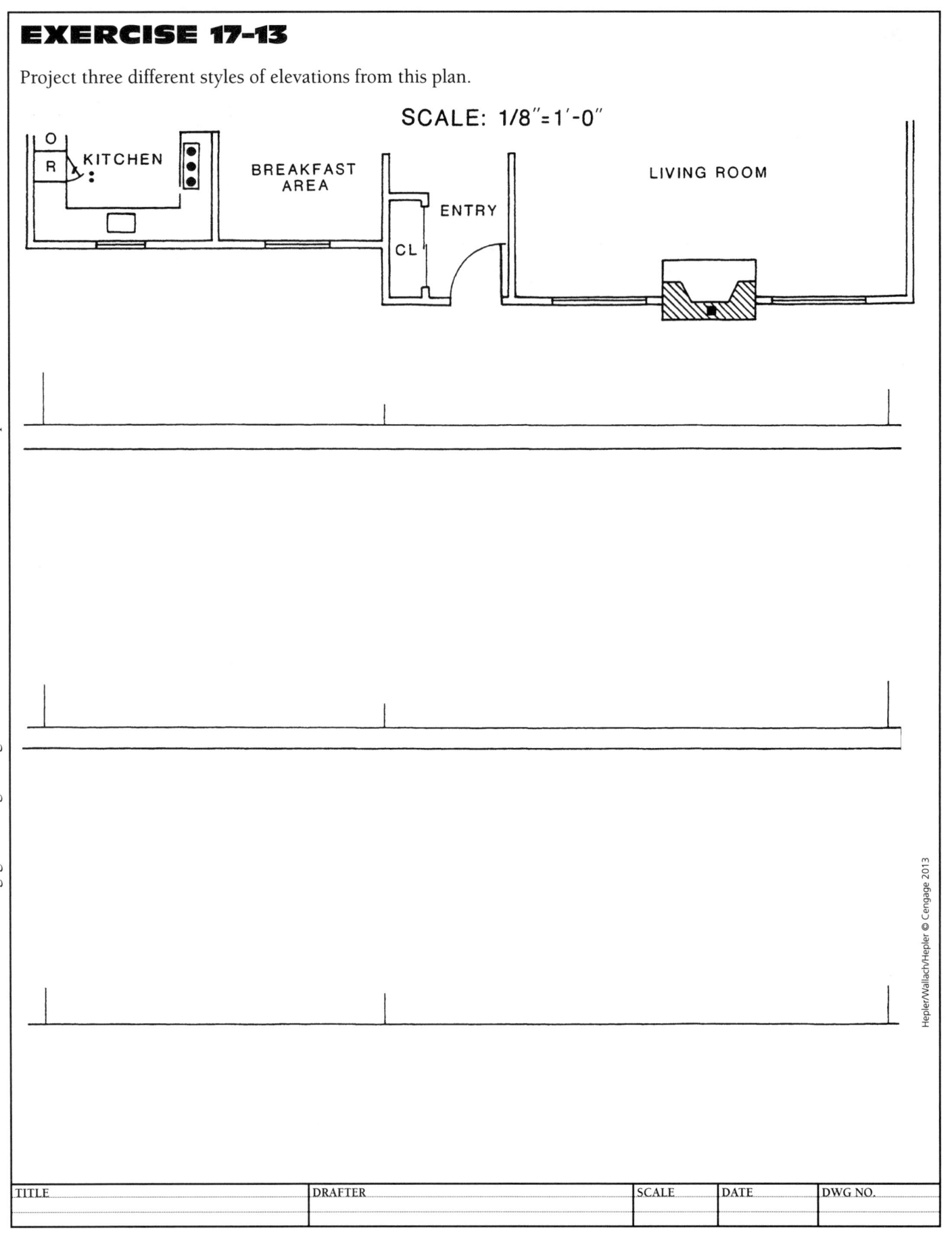

EXERCISE 17-13

Project three different styles of elevations from this plan.

SCALE: 1/8"=1'-0"

TITLE	DRAFTER	SCALE	DATE	DWG NO.

EXERCISE 17-14

From the bottom plan, project and draw an elevation view below.

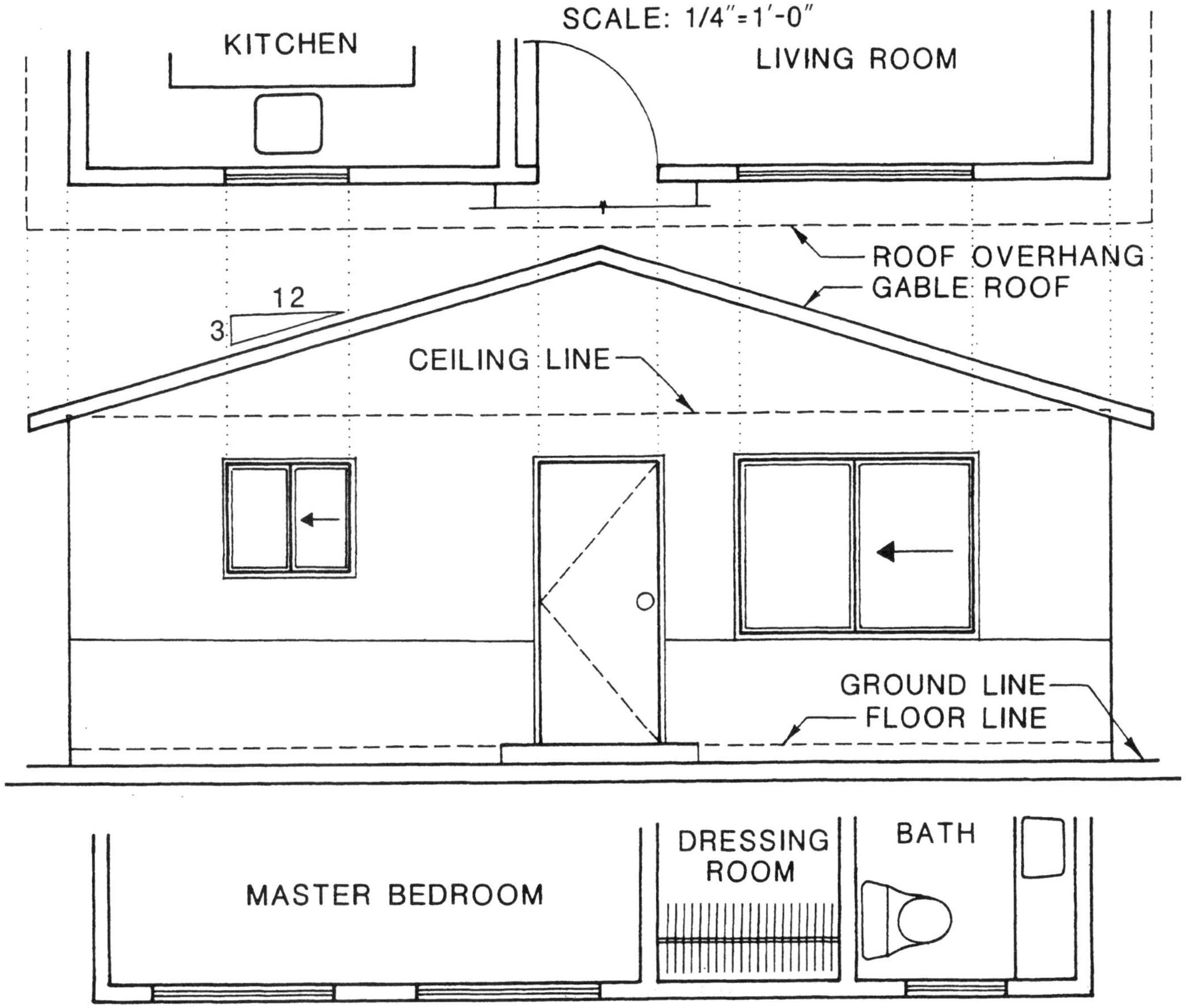
SCALE: 1/4"=1'-0"
KITCHEN
LIVING ROOM
ROOF OVERHANG
GABLE ROOF
12
3
CEILING LINE
GROUND LINE
FLOOR LINE
DRESSING ROOM
BATH
MASTER BEDROOM

Hepler/Wallach/Hepler © Cengage 2013

TITLE	DRAFTER	SCALE	DATE	DWG NO.

EXERCISE 17-15

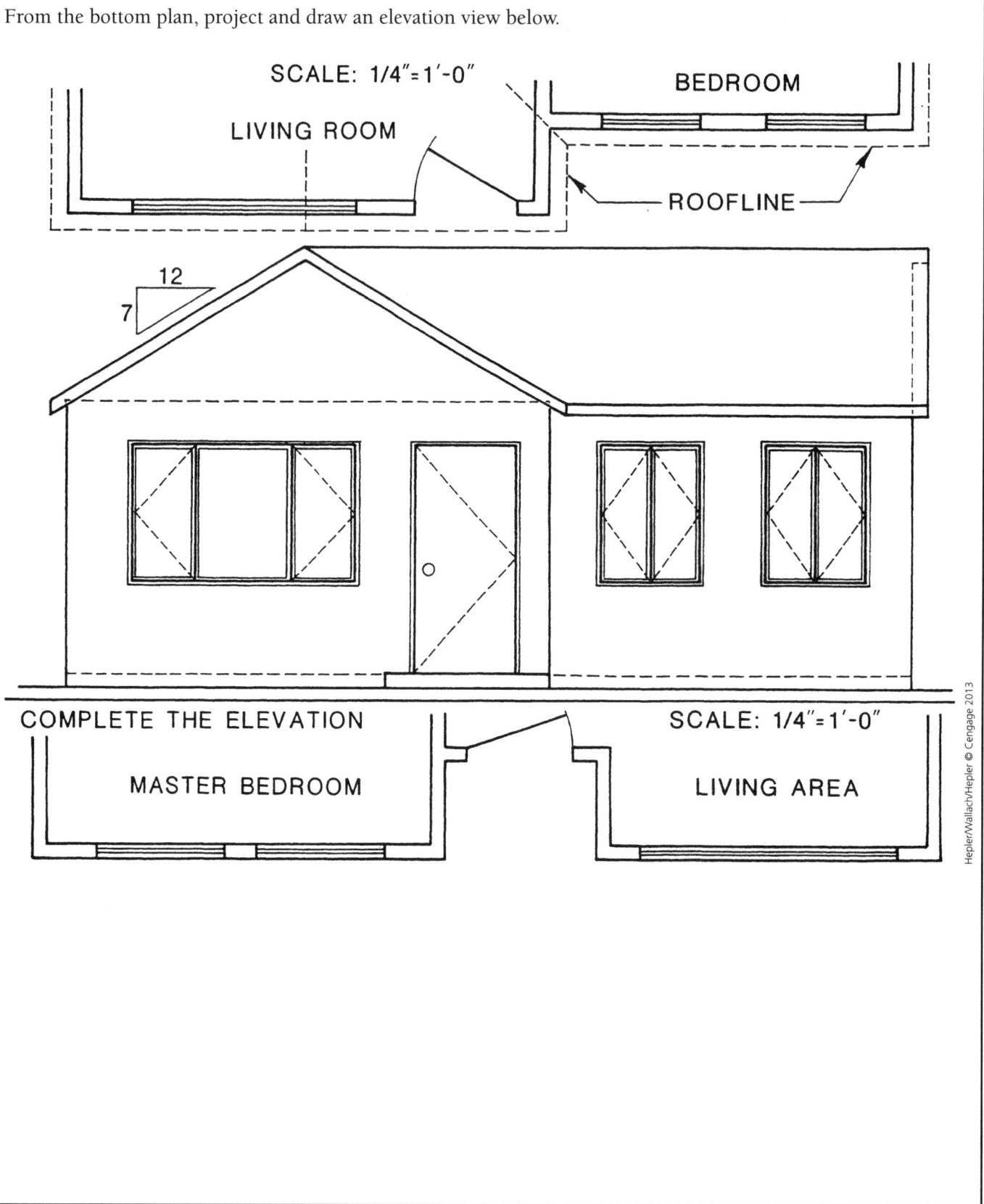

From the bottom plan, project and draw an elevation view below.

SCALE: 1/4"=1'-0"

LIVING ROOM

BEDROOM

ROOFLINE

12
7

COMPLETE THE ELEVATION

MASTER BEDROOM

SCALE: 1/4"=1'-0"

LIVING AREA

TITLE	DRAFTER	SCALE	DATE	DWG NO.

… **EXERCISE 18** Sectional Detail and Cabinetry Drawings ■ **93**

EXERCISE 18-1

Retaining the basic style (such as split level), redesign each house by making a new section drawing.

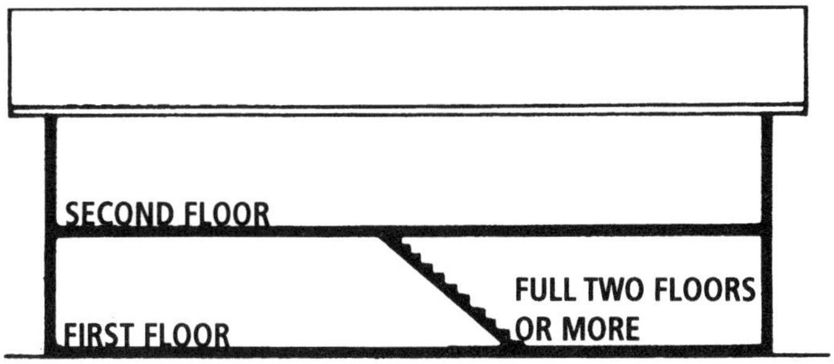

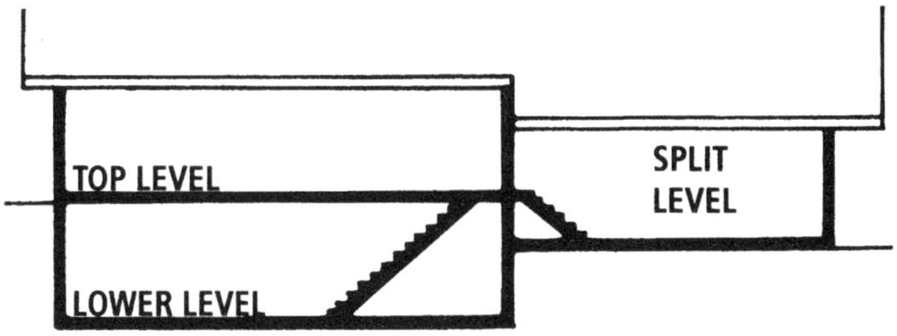

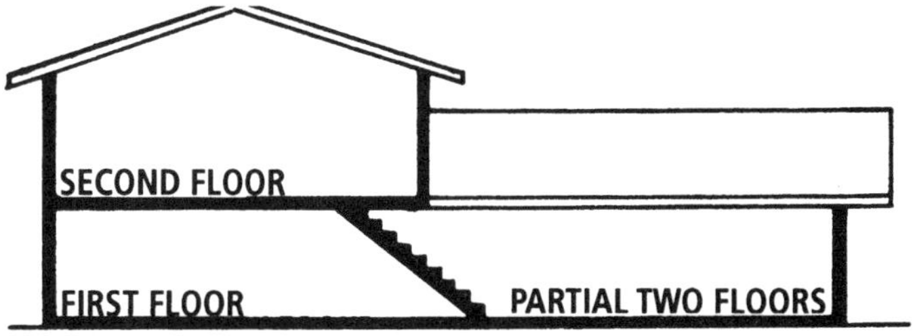

SECOND FLOOR
FIRST FLOOR
FULL TWO FLOORS OR MORE

TOP LEVEL
LOWER LEVEL
SPLIT LEVEL

SECOND FLOOR
FIRST FLOOR
PARTIAL TWO FLOORS

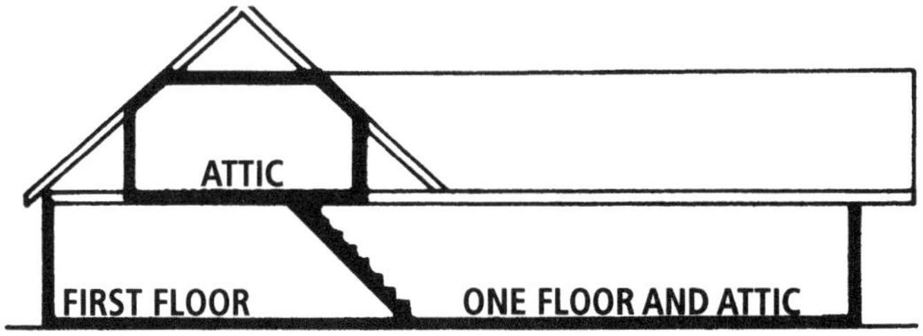

ATTIC
FIRST FLOOR
ONE FLOOR AND ATTIC

FIRST FLOOR
ONE FLOOR AND BASEMENT
BASEMENT

Hepler/Wallach/Hepler © Cengage 2013

© Cengage Learning. All rights reserved. No distribution allowed without express authorization.

TITLE	DRAFTER	SCALE	DATE	DWG NO.

EXERCISE 18-2

Draw a full section of the door, head, and sill.

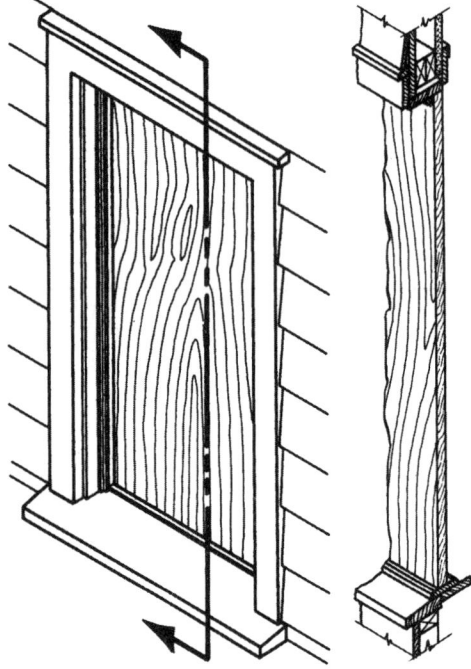

| TITLE | DRAFTER | SCALE | DATE | DWG NO. |

EXERCISE 19 Pictorial Drawings ■ 95

EXERCISE 19-1

Using the grid, make an exterior perspective drawing of a building of your own design.

TITLE	DRAFTER	SCALE	DATE	DWG NO.

EXERCISE 19-2

Using the floor plan and picture as guides, complete and render the pictorial drawing. However, change the style of the house from colonial to contemporary.

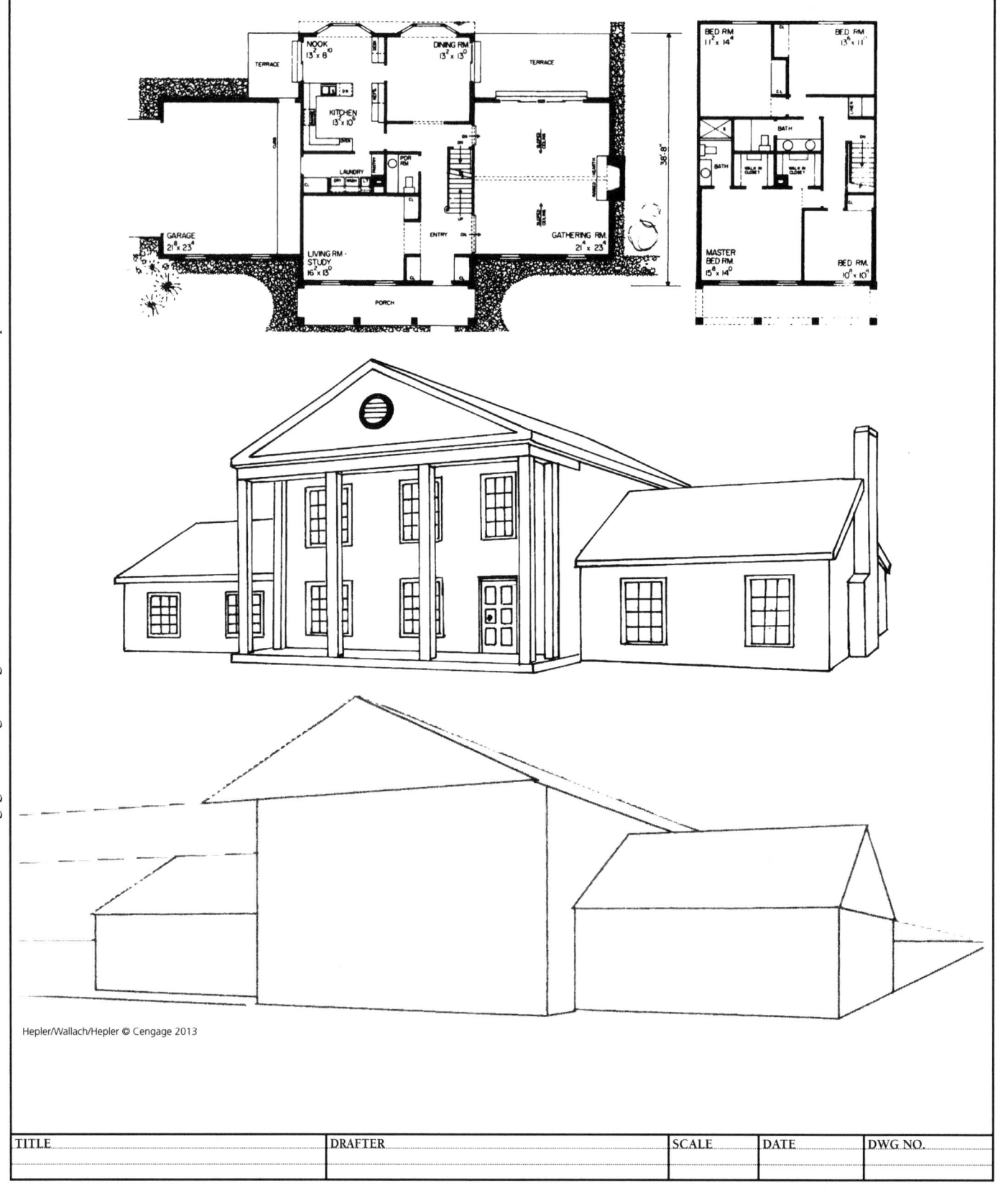

Hepler/Wallach/Hepler © Cengage 2013

TITLE	DRAFTER	SCALE	DATE	DWG NO.

EXERCISE 19-3

Using the grid, draw a one-point, interior perspective of a room of your own design.

| TITLE | DRAFTER | SCALE | DATE | DWG NO. |

EXERCISE 19-4

Using the grid, draw a two-point, interior perspective of a room of your own design.

TITLE	DRAFTER	SCALE	DATE	DWG NO.

EXERCISE 19-5

Complete the two-point interior perspective of a room of your choice.

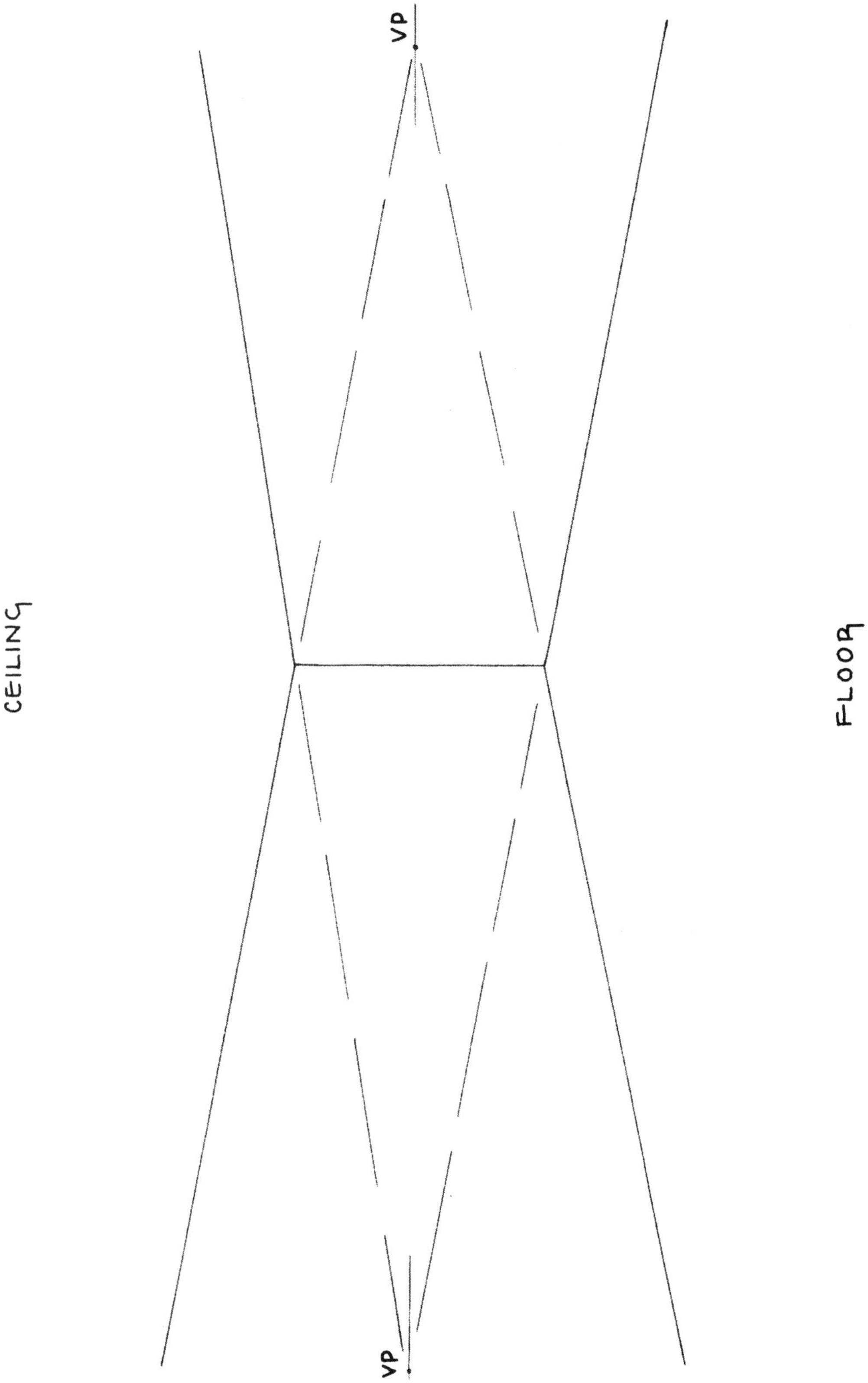

CEILING

FLOOR

VP

VP

TITLE	DRAFTER	SCALE	DATE	DWG NO.

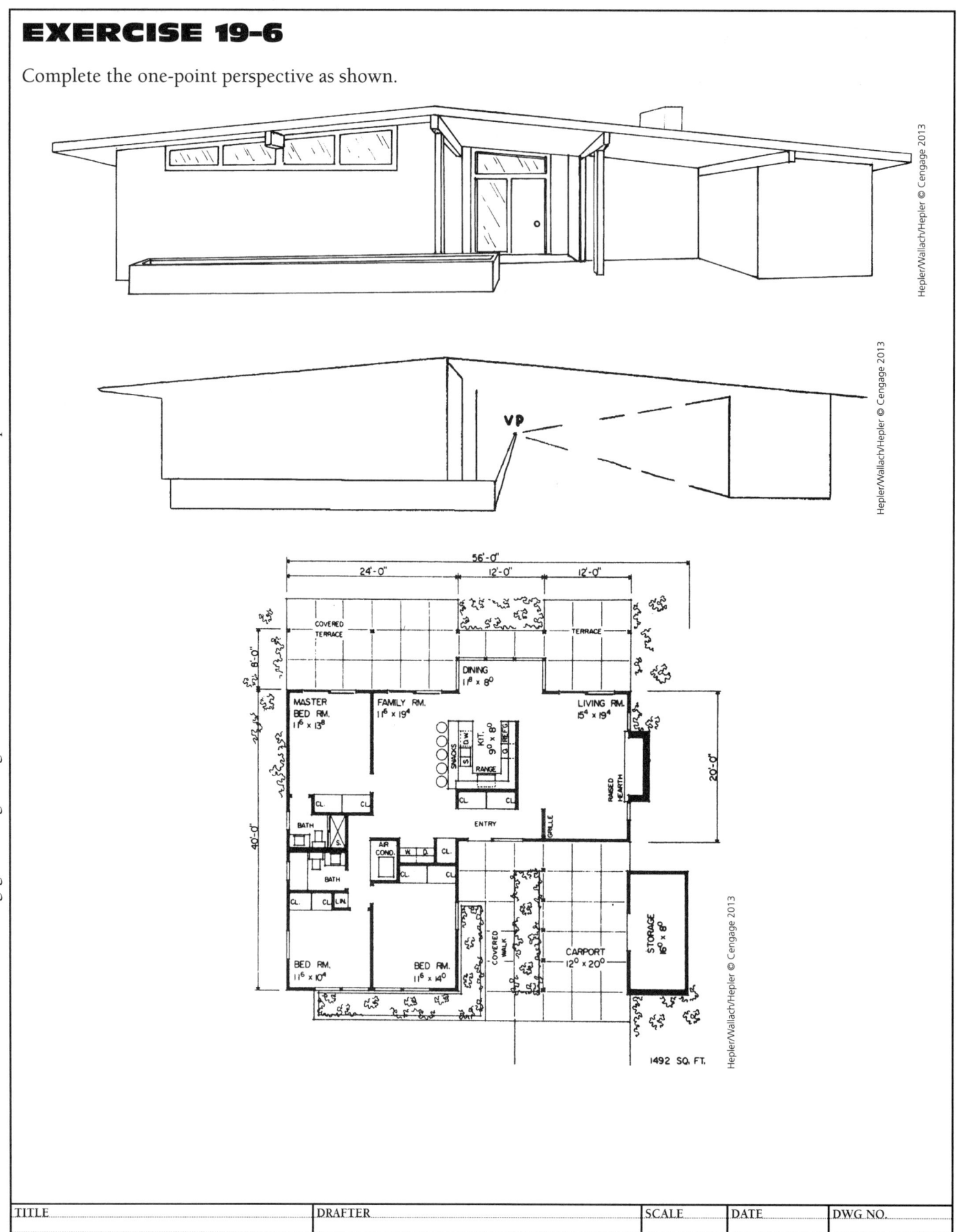

EXERCISE 19-6

Complete the one-point perspective as shown.

EXERCISE 19-7

Complete the two-point perspective as shown.

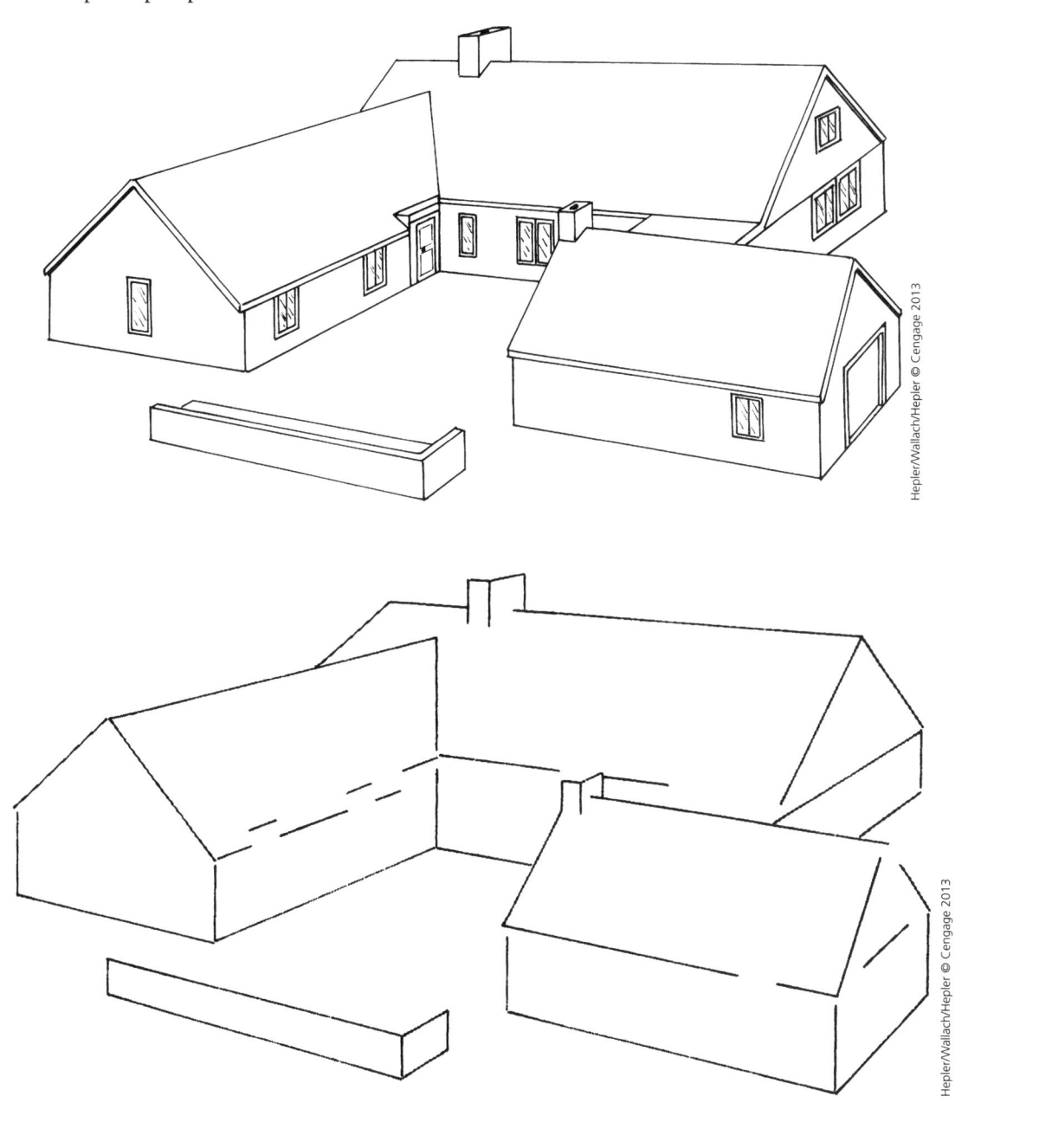

TITLE	DRAFTER	SCALE	DATE	DWG NO.

EXERCISE 19-8

Complete the two-point exterior perspective as shown.

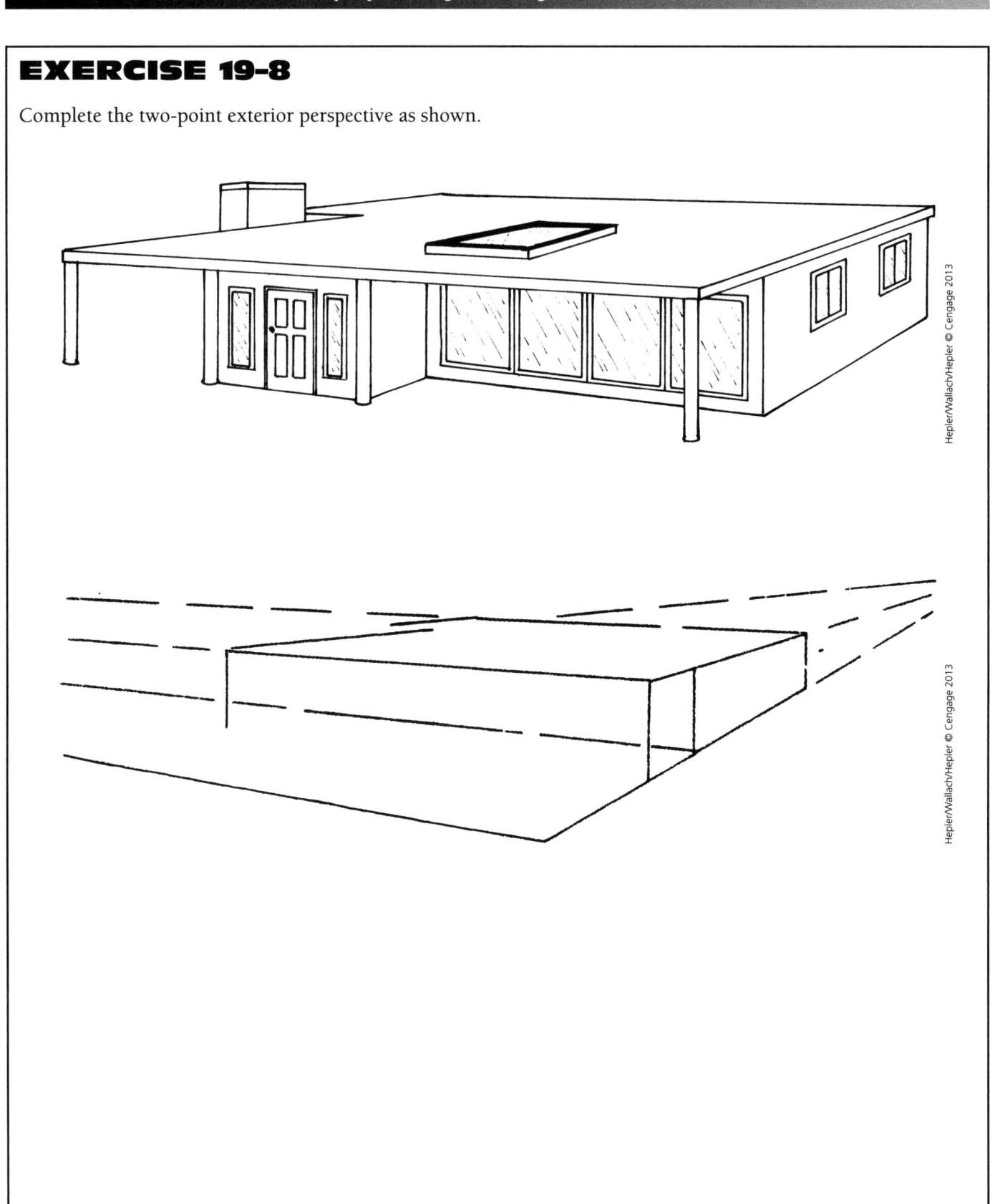

TITLE	DRAFTER	SCALE	DATE	DWG NO.

EXERCISE 19-9

Complete the two-point exterior perspective as shown.

| TITLE | DRAFTER | SCALE | DATE | DWG NO. |

EXERCISE 19-10

Multiple-choice questions: Select the correct answer and write the letter in the space at the right.

1. A presentation drawing is used for
   a. showing the construction procedures.
   b. a set of working drawings.
   c. instruction on perspective drawings.
   d. selling the appearance of the structure.

   1. _____

2. A drawing with a high perceptive view will emphasize the
   a. roof.
   b. center of walls.
   c. foundation/ground.
   d. none of the above.

   2. _____

3. A drawing with a low perspective view will emphasize the
   a. roof.
   b. center of walls.
   c. foundation/ground.
   d. none of the above.

   3. _____

4. The type of drawing that is not usually used as a presentation drawing is a(n)
   a. elevation.
   b. interior perspective.
   c. foundation plan.
   d. floor plan.

   4. _____

5. What is the "angle of view"?
   a. The angle at which the perspective lines meet
   b. The viewing of the perspective drawing
   c. An approximation of the distance of the viewer from the object being drawn
   d. The line of sight

   5. _____

6. An average angular position for drawing a perspective of a home is
   a. 30°.
   b. 135°.
   c. 90°.
   d. 185°.

   6. _____

Complete the one-point interior perspective of a room of your choice.

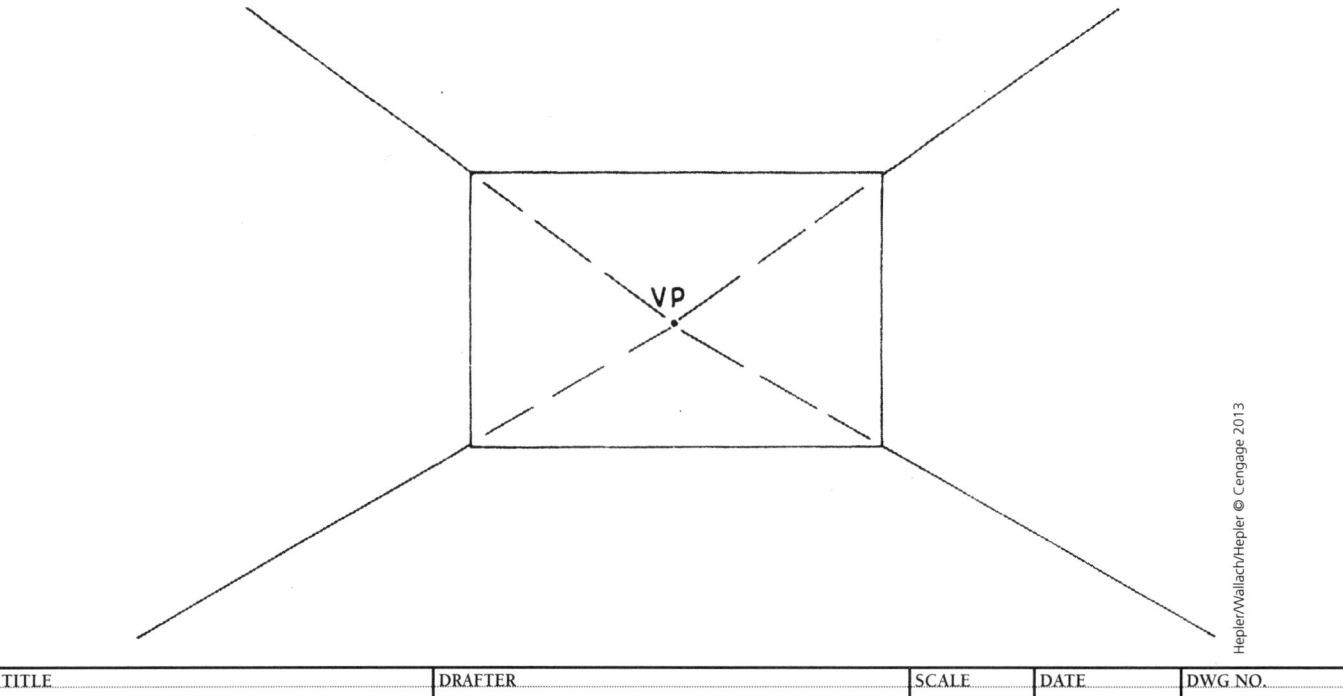

TITLE	DRAFTER	SCALE	DATE	DWG NO.

EXERCISE 20-1

Study the textures shown in each illustration. Copy the drawings in the space provided.

TITLE	DRAFTER	SCALE	DATE	DWG NO.

EXERCISE 20-2

Draw an elevation of your own design. Use the rendering techniques shown in the example drawing.

TITLE	DRAFTER	SCALE	DATE	DWG NO.

EXERCISE 20-3

Study the texturizing of the illustrations below. Duplicate each in the space provided.

TITLE	DRAFTER	SCALE	DATE	DWG NO.

EXERCISE 21-1

Cut out this drawing of a kitchen and assemble it into a 3D model. Then use the furniture models on the next page to furnish it.

KITCHEN
11'-6" x 9'-6"

EXERCISE 21-2

Cut out the items below and create 3D models. Use them to furnish the kitchen model.

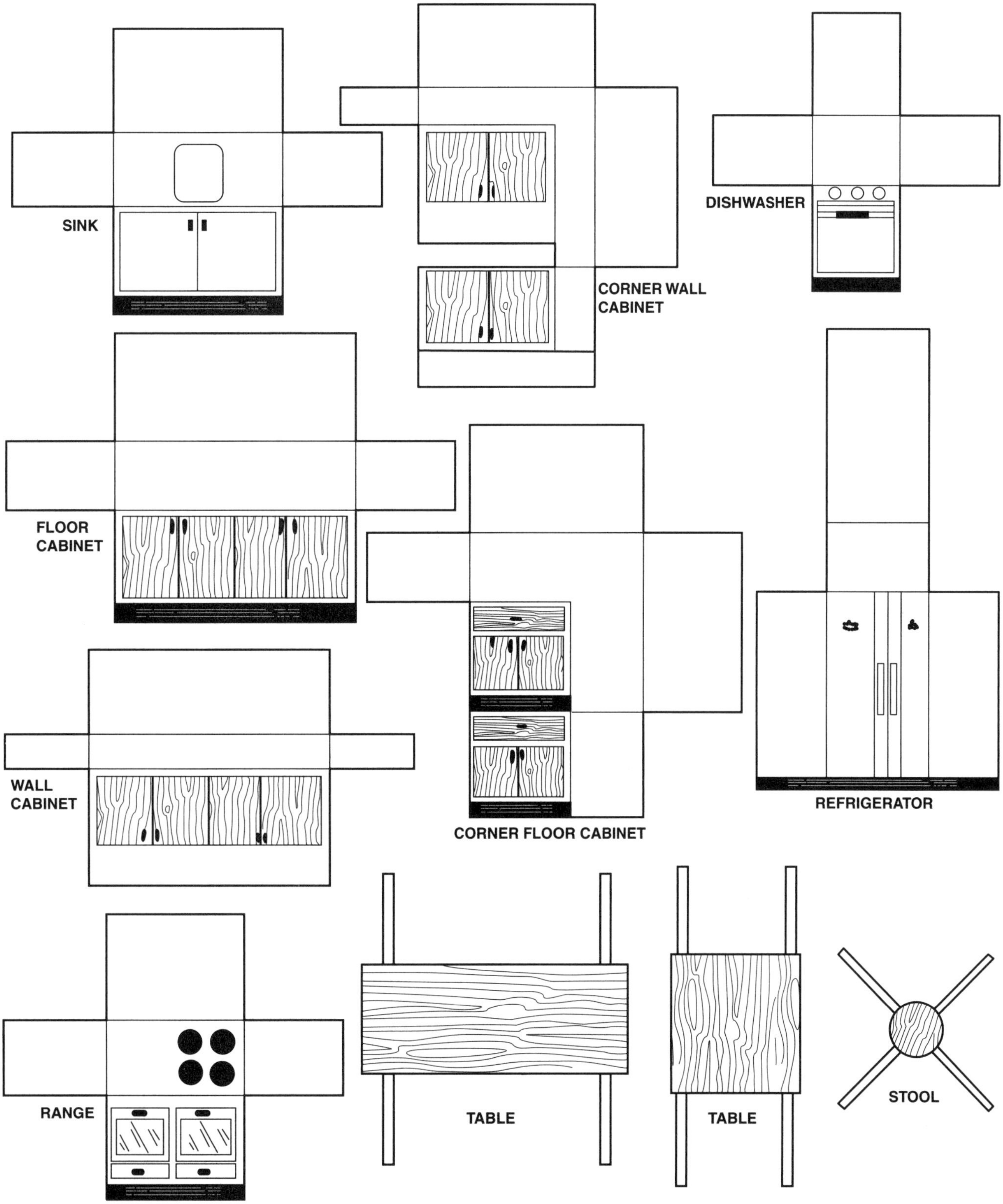

SINK

CORNER WALL CABINET

DISHWASHER

FLOOR CABINET

WALL CABINET

CORNER FLOOR CABINET

REFRIGERATOR

RANGE

TABLE

TABLE

STOOL

TITLE	DRAFTER	SCALE	DATE	DWG NO.

EXERCISE 21-3

Cut out and fold the room walls upward. Add furniture to create a 3D room model.

BEDROOM
12'-0" x 13'-0"

TITLE	DRAFTER	SCALE	DATE	DWG NO.

EXERCISE 21-4

Cut out, fold, and assemble the furniture models from pages 112 and 113 with transparent tape. Place furniture in each room. Then color in the floor and wall treatments before assembling each room. Once the design and furniture placement are established, attach the walls together with transparent tape.

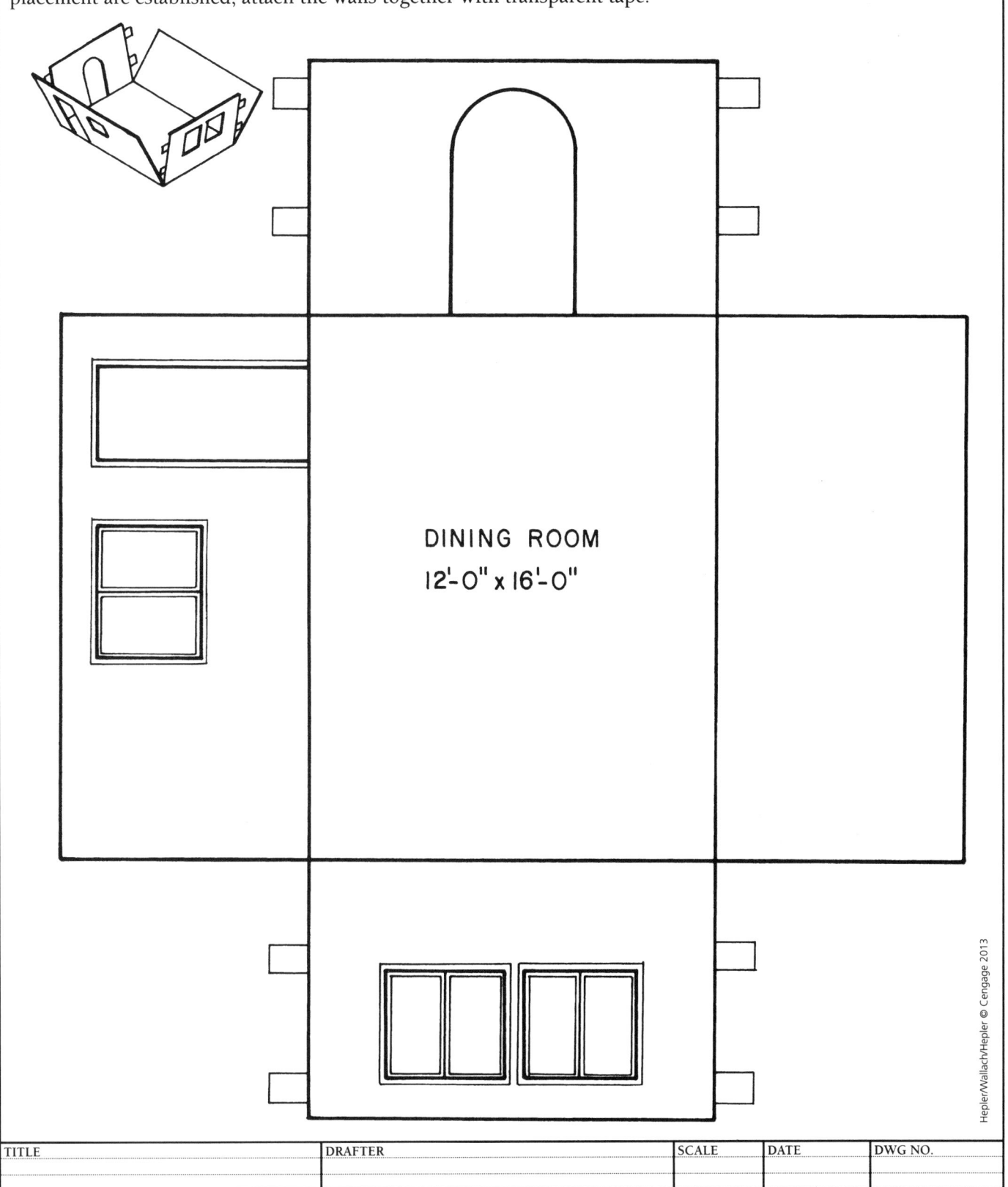

DINING ROOM
12'-0" x 16'-0"

TITLE	DRAFTER	SCALE	DATE	DWG NO.

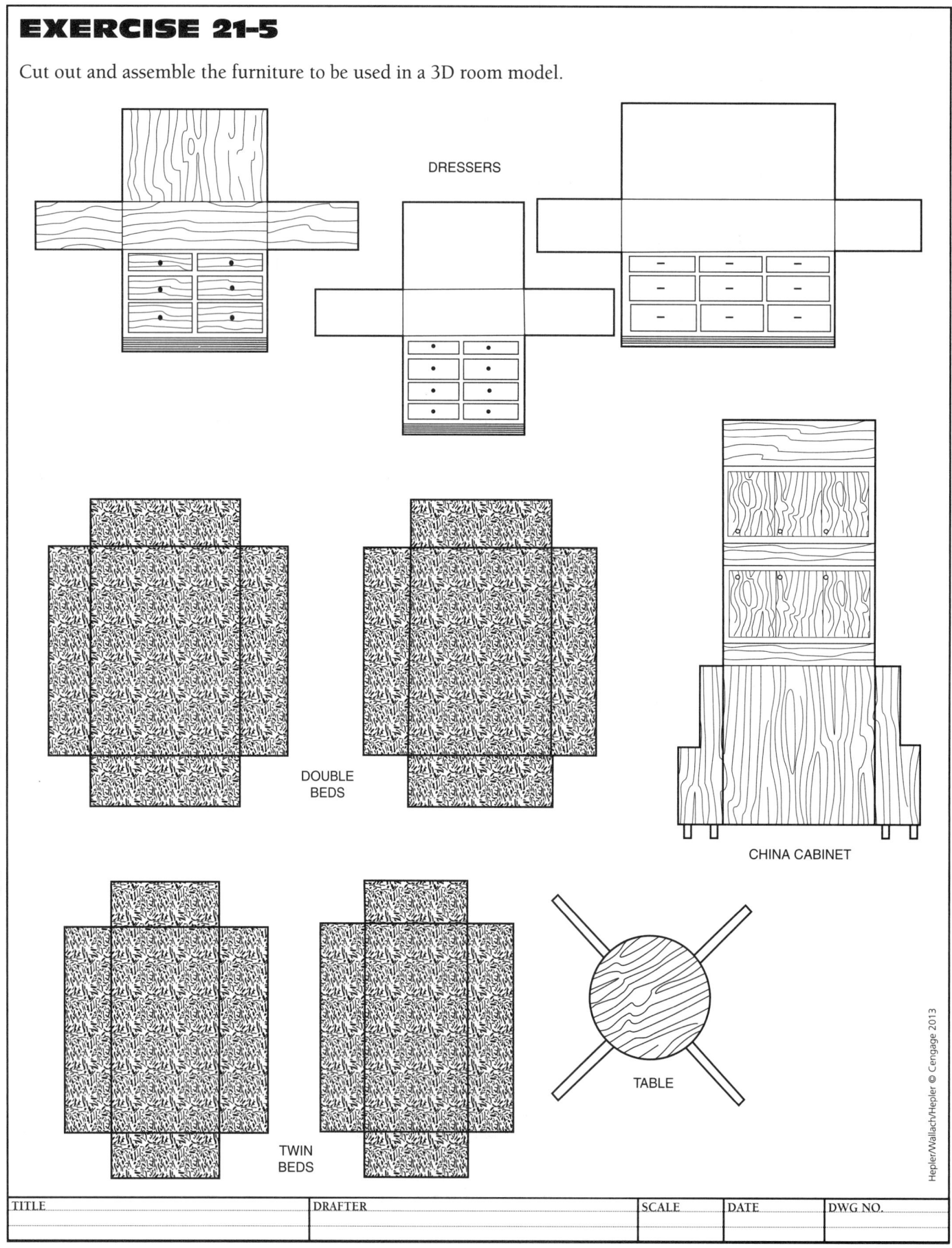

EXERCISE 21-5

Cut out and assemble the furniture to be used in a 3D room model.

DRESSERS

DOUBLE BEDS

CHINA CABINET

TWIN BEDS

TABLE

TITLE	DRAFTER	SCALE	DATE	DWG NO.

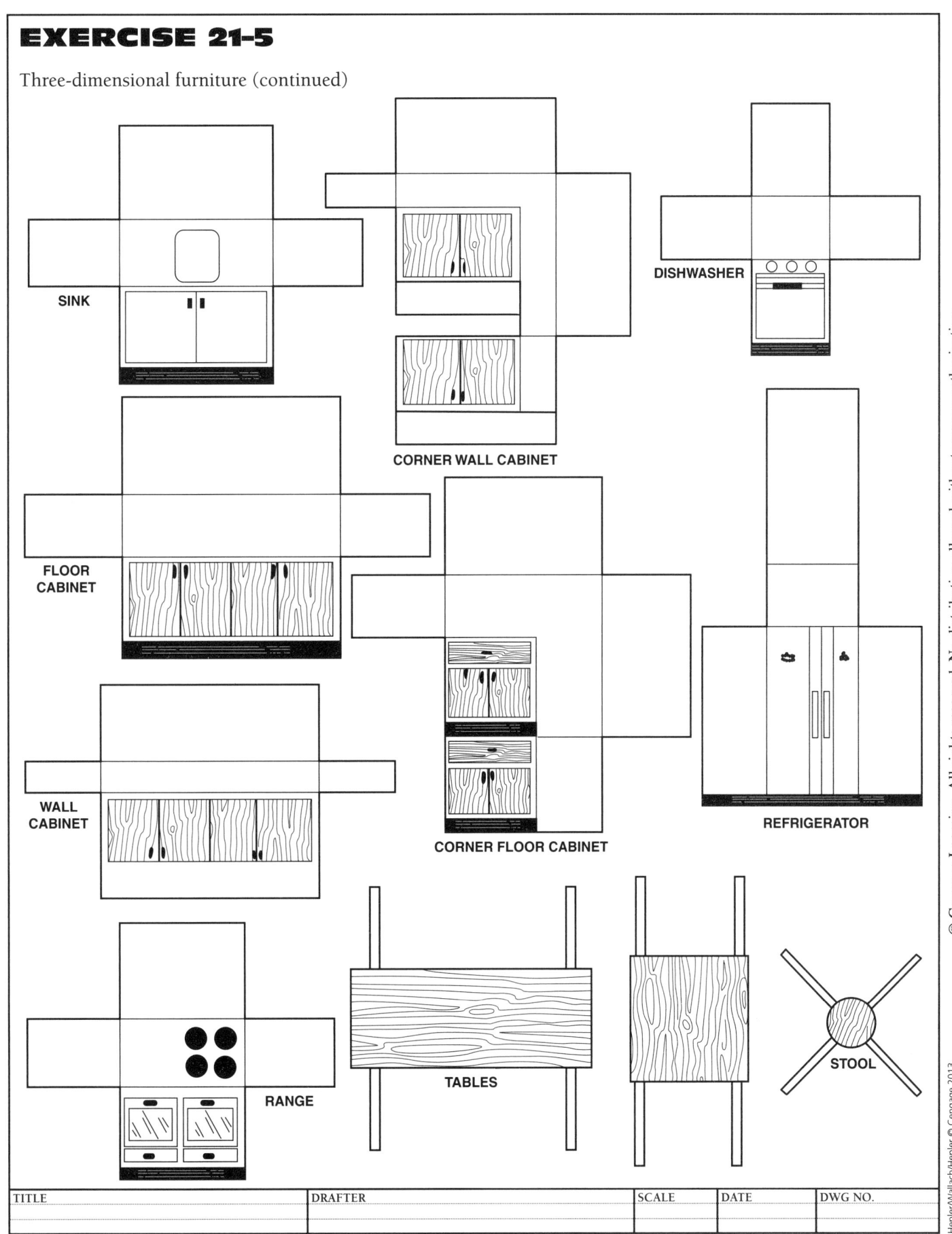

EXERCISE 21-5

Three-dimensional furniture (continued)

SINK

CORNER WALL CABINET

DISHWASHER

FLOOR CABINET

WALL CABINET

CORNER FLOOR CABINET

REFRIGERATOR

RANGE

TABLES

STOOL

TITLE	DRAFTER	SCALE	DATE	DWG NO.

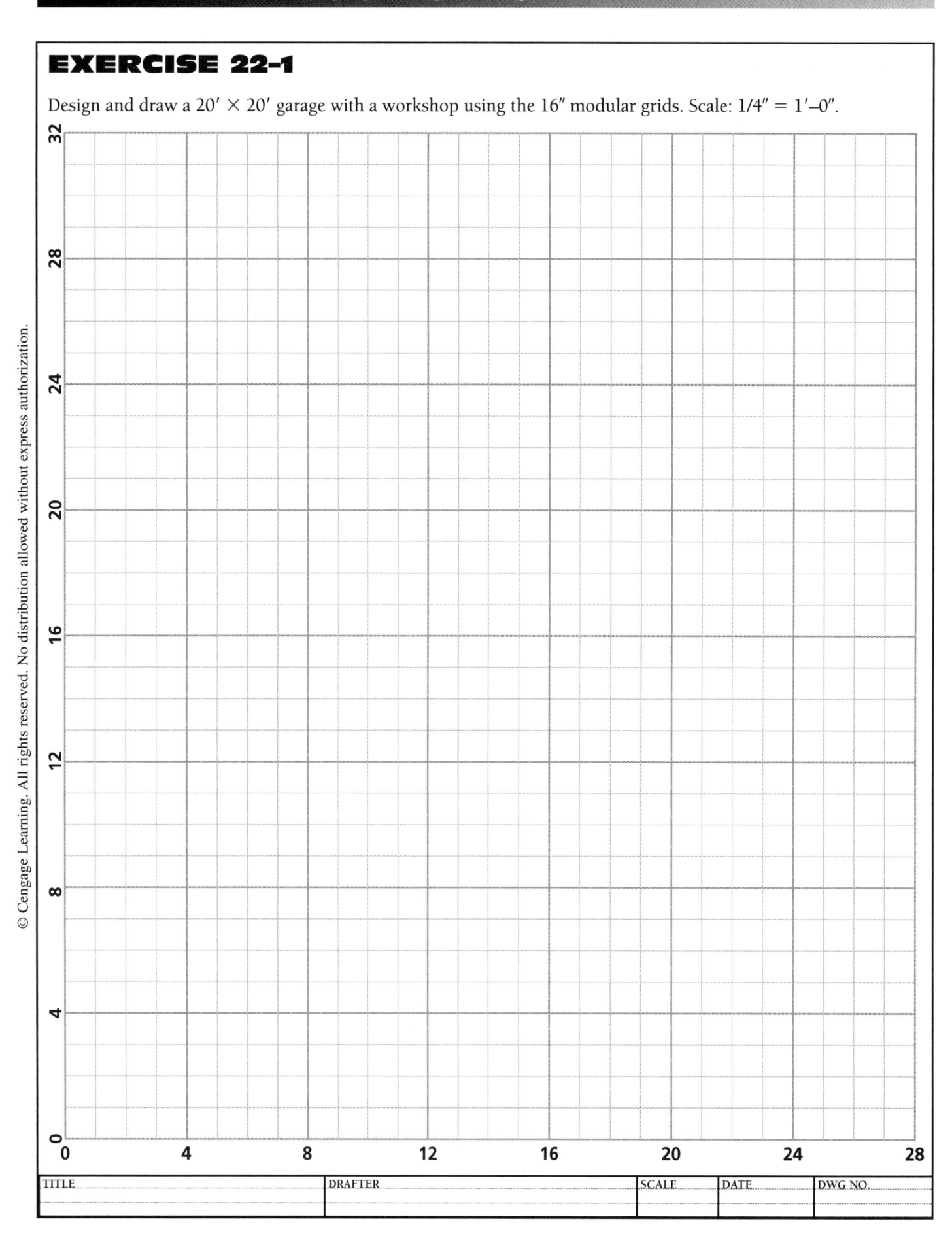

EXERCISE 22-1

Design and draw a 20′ × 20′ garage with a workshop using the 16″ modular grids. Scale: 1/4″ = 1′–0″.

| TITLE | DRAFTER | SCALE | DATE | DWG NO. |

ized.
EXERCISE 22-2

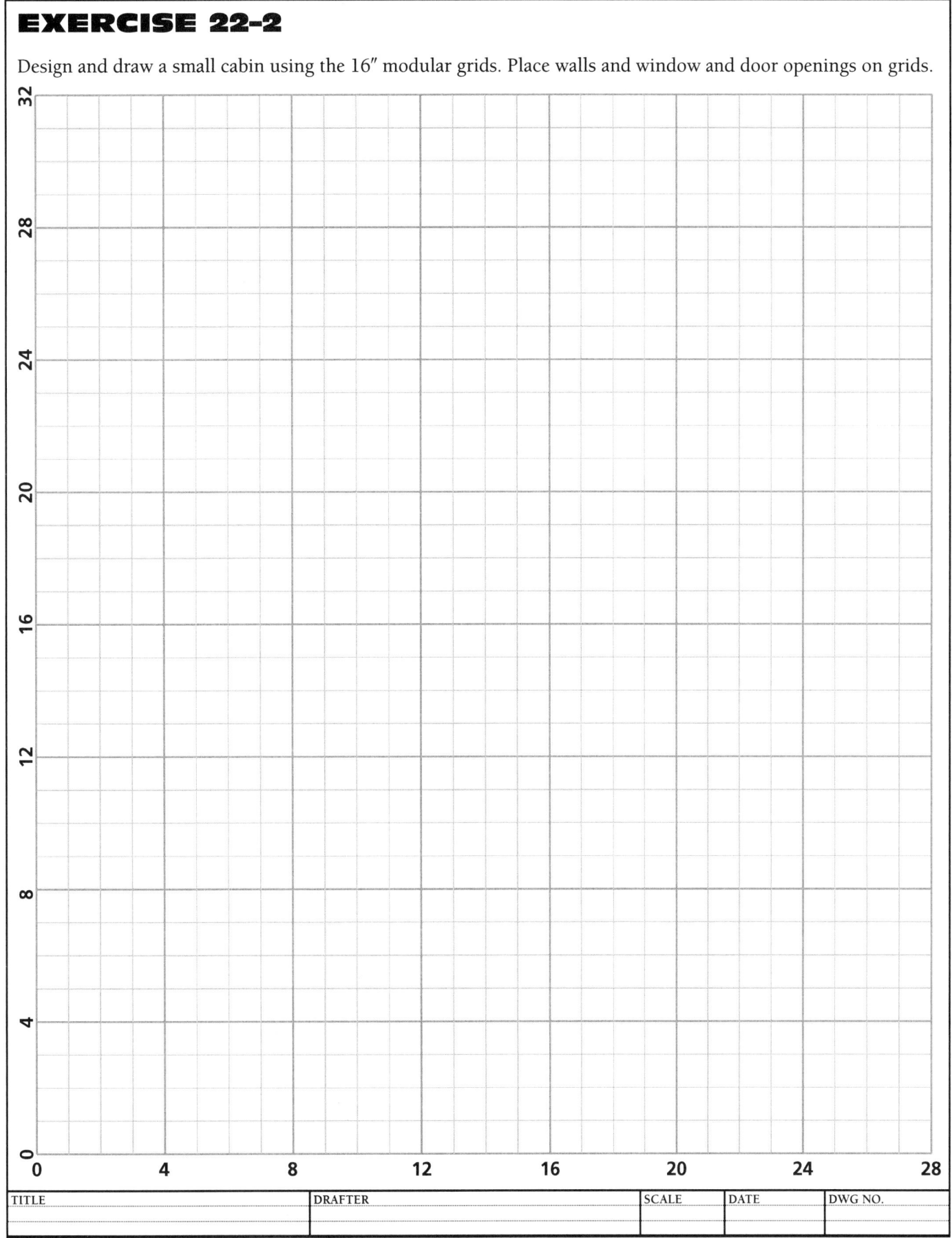

Design and draw a small cabin using the 16″ modular grids. Place walls and window and door openings on grids.

TITLE	DRAFTER	SCALE	DATE	DWG NO.

EXERCISE 22-3

Design and draw a plan view of this garage and apartment in the space below.

TITLE	DRAFTER	SCALE	DATE	DWG NO.

EXERCISE 23-1

Complete these T- and slab-foundation plans for a single-level structure. Use the structural codes in Chapter 23 of your text for girder and joist sizes, spans, and spacing.

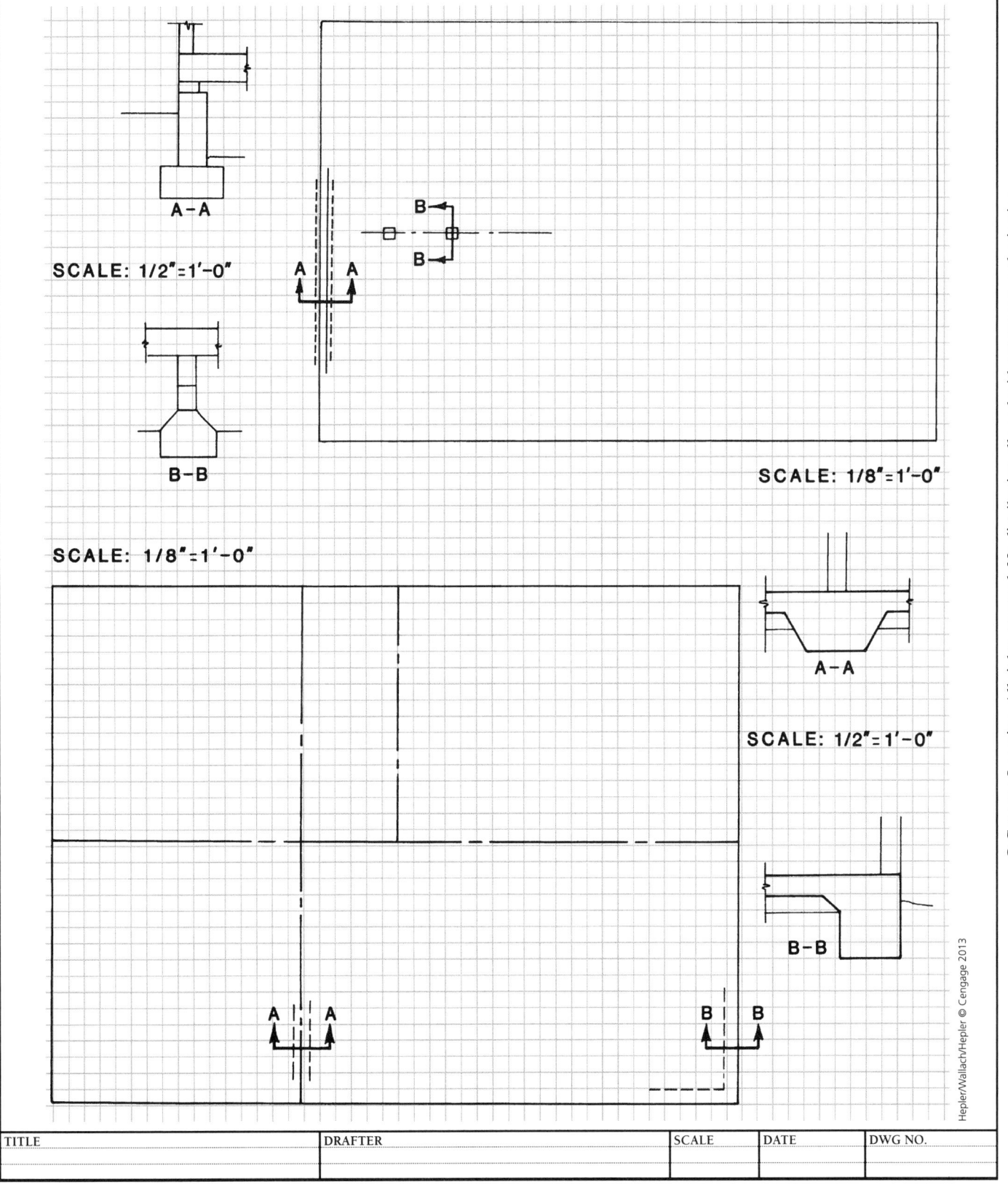

A–A
SCALE: 1/2"=1'-0"

B–B
SCALE: 1/8"=1'-0"

SCALE: 1/8"=1'-0"

A–A
SCALE: 1/2"=1'-0"

B–B

TITLE	DRAFTER	SCALE	DATE	DWG NO.

EXERCISE 23-2

Redraw this fireplace section. The scale is 3/8″ = 1′–0″.

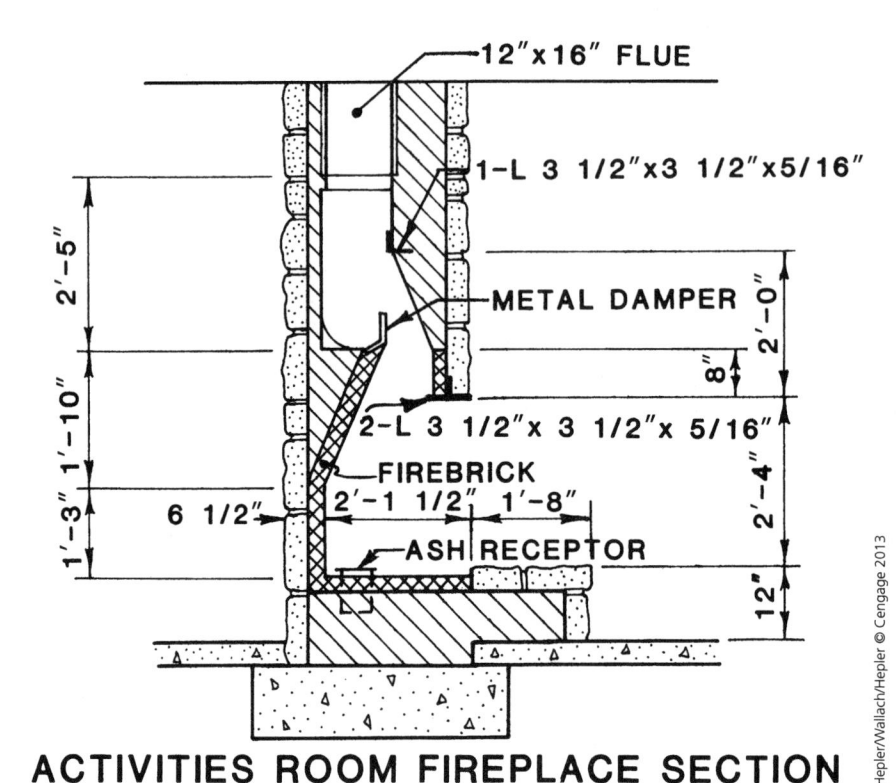

ACTIVITIES ROOM FIREPLACE SECTION

TITLE	DRAFTER	SCALE	DATE	DWG NO.

EXERCISE 23-5

Complete the T-foundation plan below using local building codes to determine member sizes, spacing, and spans.

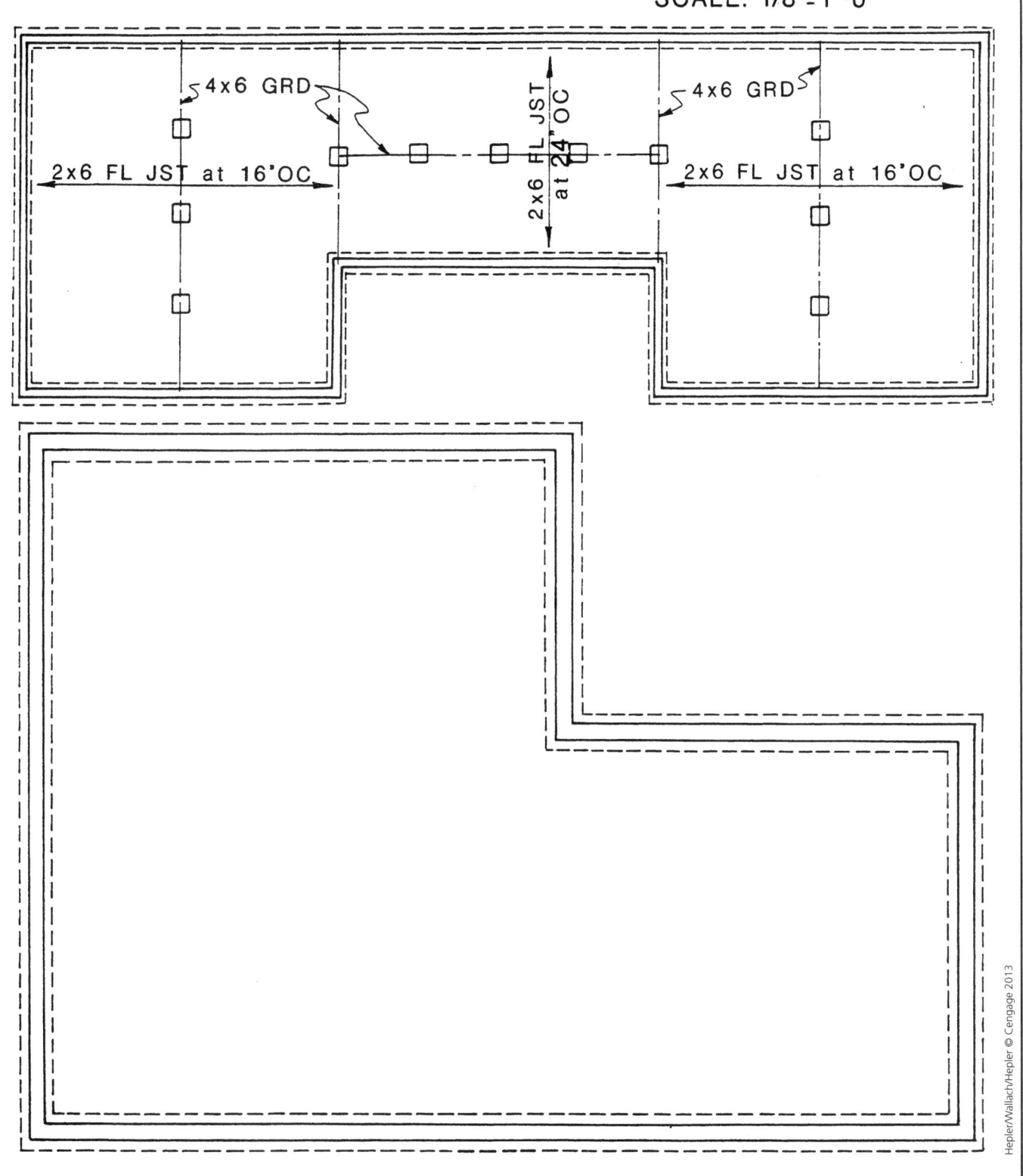

SCALE: 1/8"=1'-0"

- 4x6 GRD
- 2x6 FL JST at 16'OC
- 2x6 FL JST at 24"OC
- 4x6 GRD
- 2x6 FL JST at 16'OC

| TITLE | DRAFTER | SCALE | DATE | DWG NO. |

EXERCISE 23-6

Using the top drawing as reference, design a floor plan and draw a foundation plan for a 30′ × 50′ cabin below.

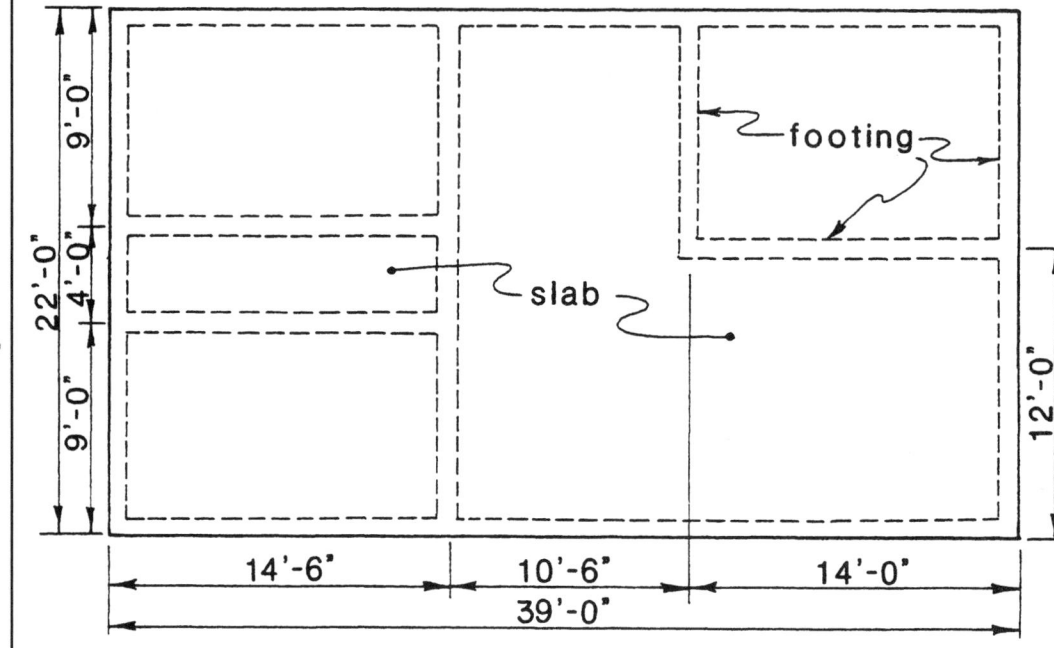

SCALE: 1/8″=1′-0″

Dimensions: 22′-0″ overall height (9′-0″, 4′-0″, 9′-0″); 12′-0″; 14′-6″, 10′-6″, 14′-0″; 39′-0″ overall.

Labels: footing, slab

SLAB CONSTRUCTION DATA

* Footings 12″ wide
* Slab 4″ to 5″ thick
* Interior footings centered with interior walls
* Interior footings for bearing walls only
* Check building codes for amount of steel reinforcement
* Slab height above finish grade is 6″ to 8″
* Provide vapor barrier under slab

TITLE	DRAFTER	SCALE	DATE	DWG NO.

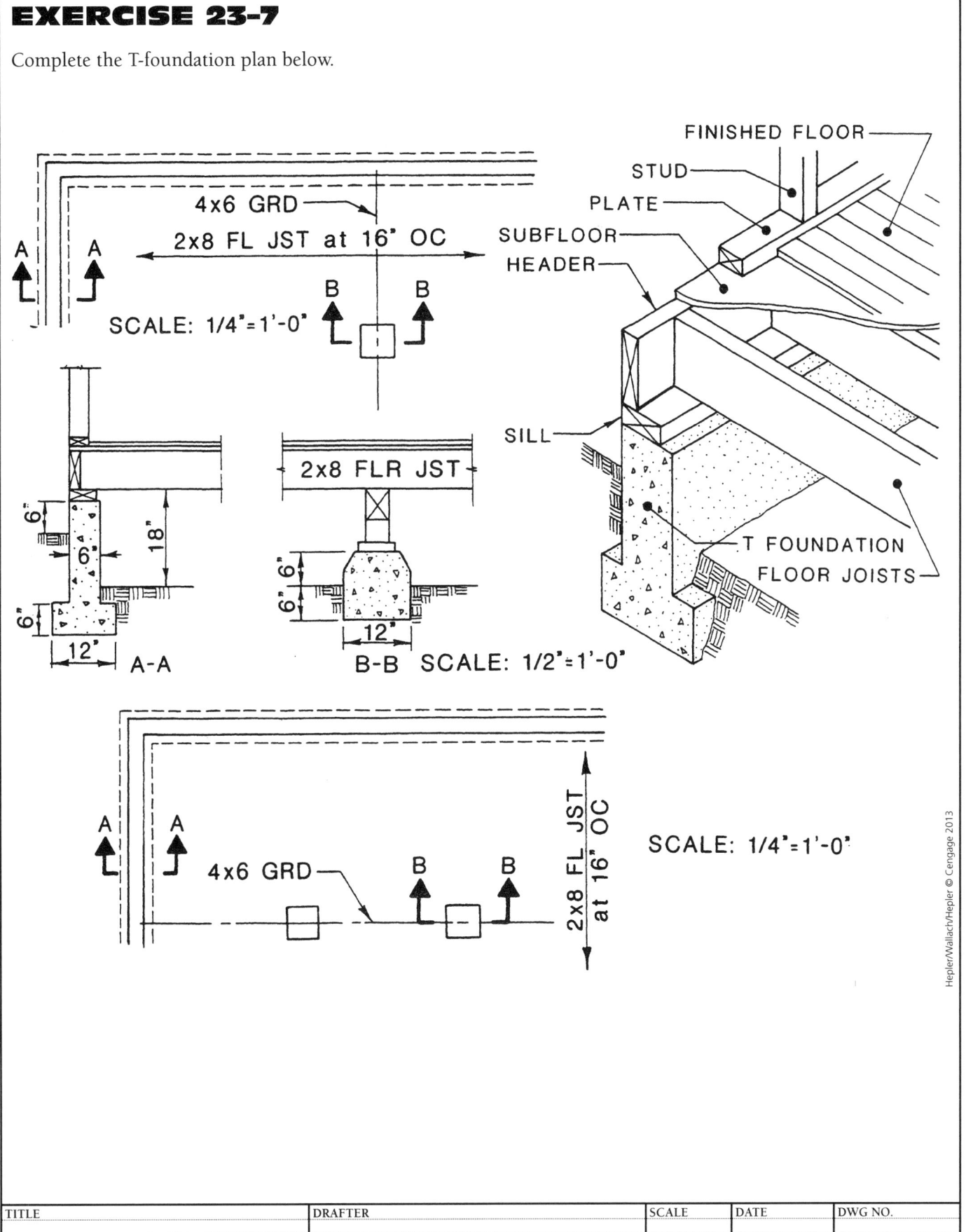

EXERCISE 23-7

Complete the T-foundation plan below.

EXERCISE 23-8

Draw a T-foundation wall section for this plan.

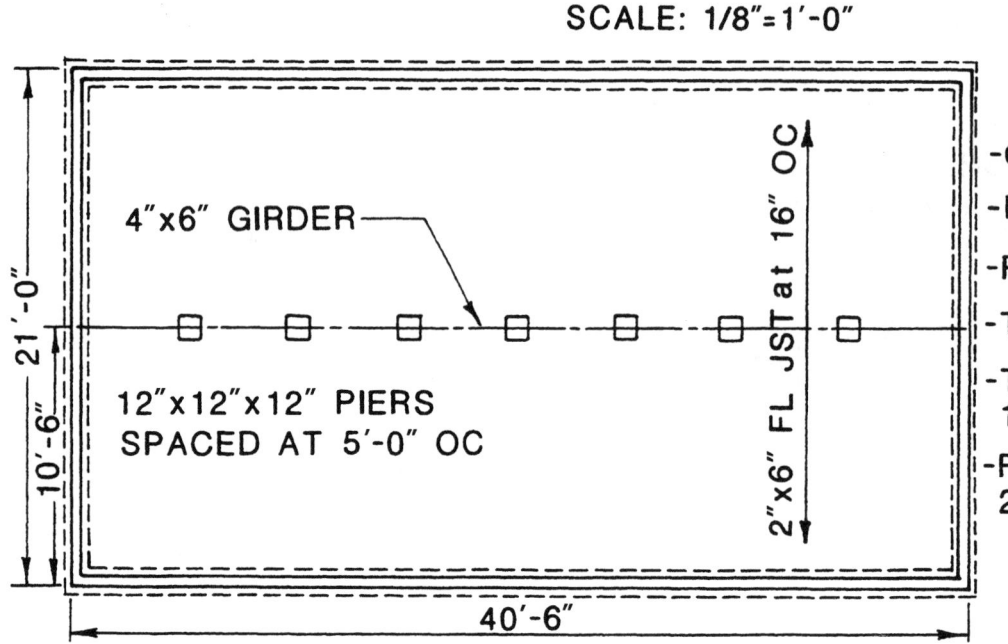

SCALE: 1/8"=1'-0"

- 4"x6" GIRDER
- 12"x12"x12" PIERS SPACED AT 5'-0" OC
- 2"x6" FL JST at 16" OC
- 21'-0"
- 10'-6"
- 40'-6"

GENERAL T FOUNDATION DESIGN DATA

- GIRDER 4"x6"
- PIERS 12"x12"x12"
- PIERS SPACED AT 5"
- T WALL WIDTH 6"
- T FOOTING WIDTH 12"
- FLOOR JOISTS 2"x6" SPACED AT 16"

TITLE	DRAFTER	SCALE	DATE	DWG NO.

EXERCISE 23-9

Refer to sections A–A and B–B and complete the slab-foundation plan below. Scale is 1/8" = 1'–0".

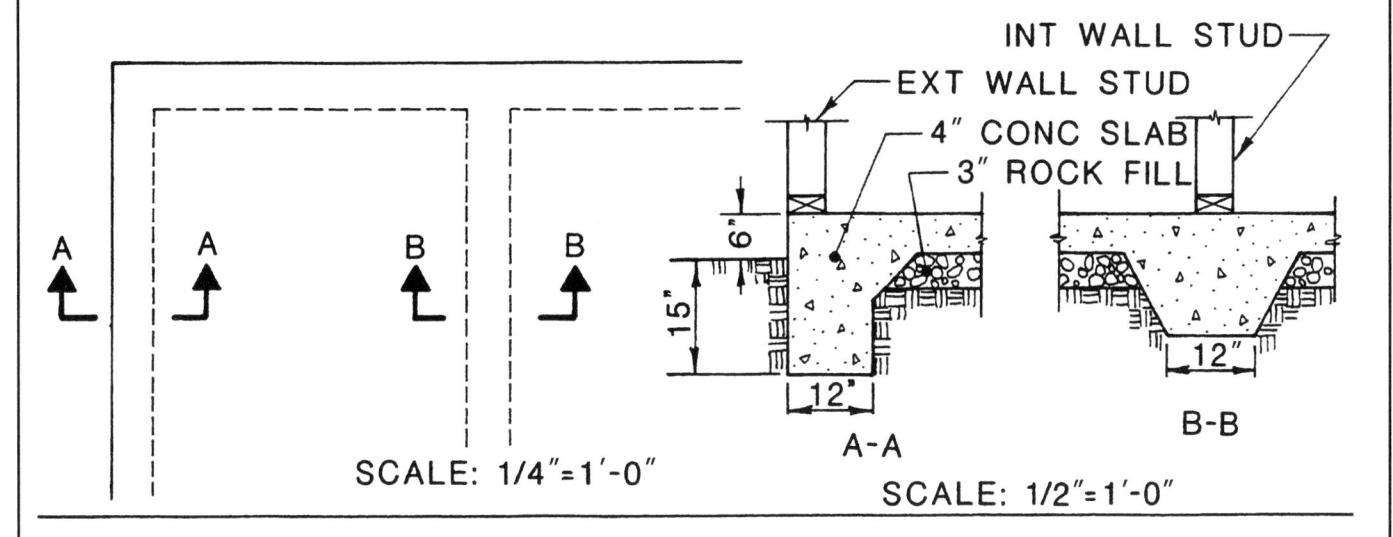

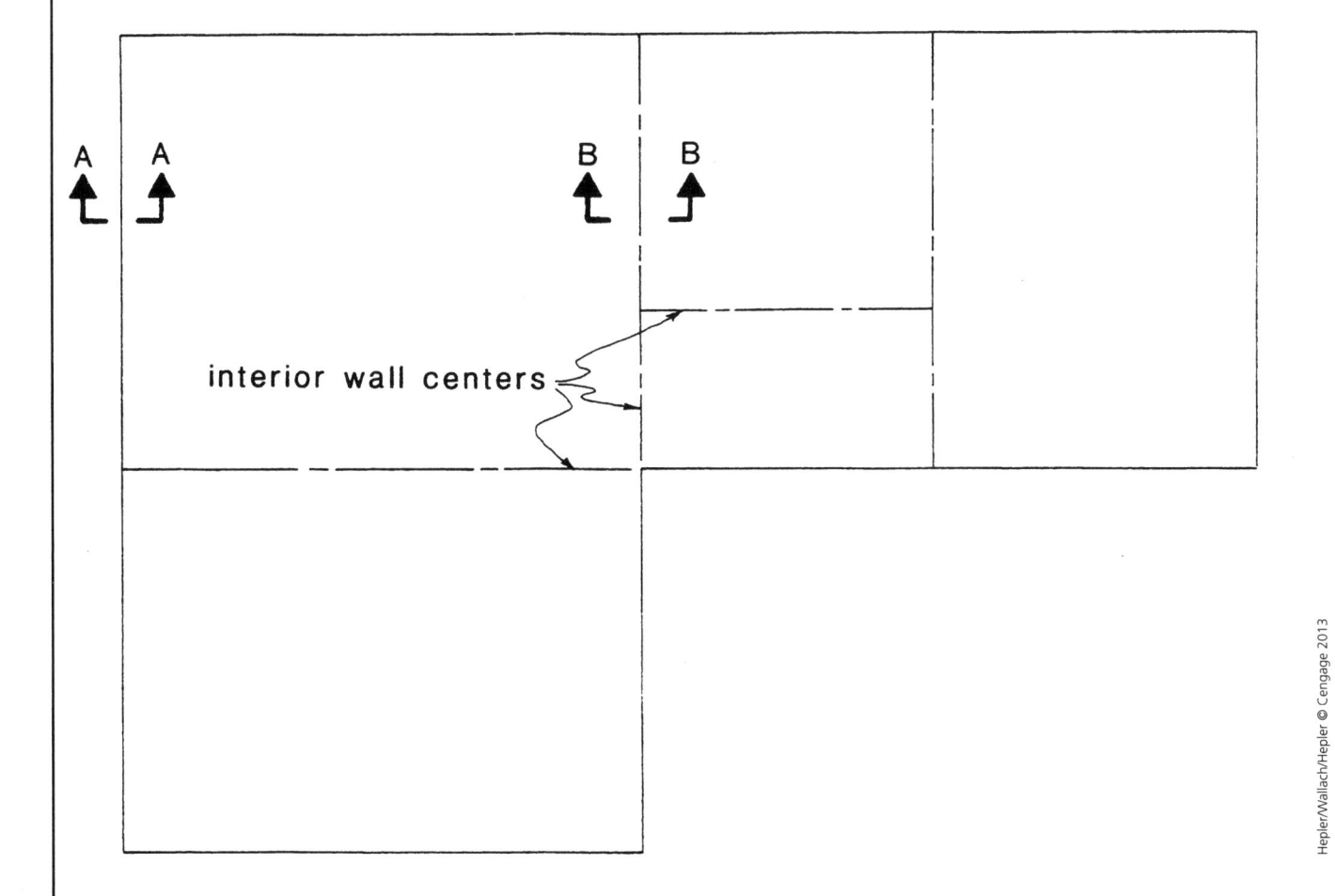

TITLE	DRAFTER	SCALE	DATE	DWG NO.

EXERCISE 24-1

Redraw this wood-frame system using the isometric grid provided.

- DOUBLE TOP PLATE
- LET-IN DIAGONAL BRACE
- STUD
- SOLE PLATE
- DIAGONAL SUBFLOOR
- HEADER
- SUBFLOOR
- SILL
- T FOUNDATION
- CORNER POST
- FILLER BLOCK

| TITLE | DRAFTER | SCALE | DATE | DWG NO. |

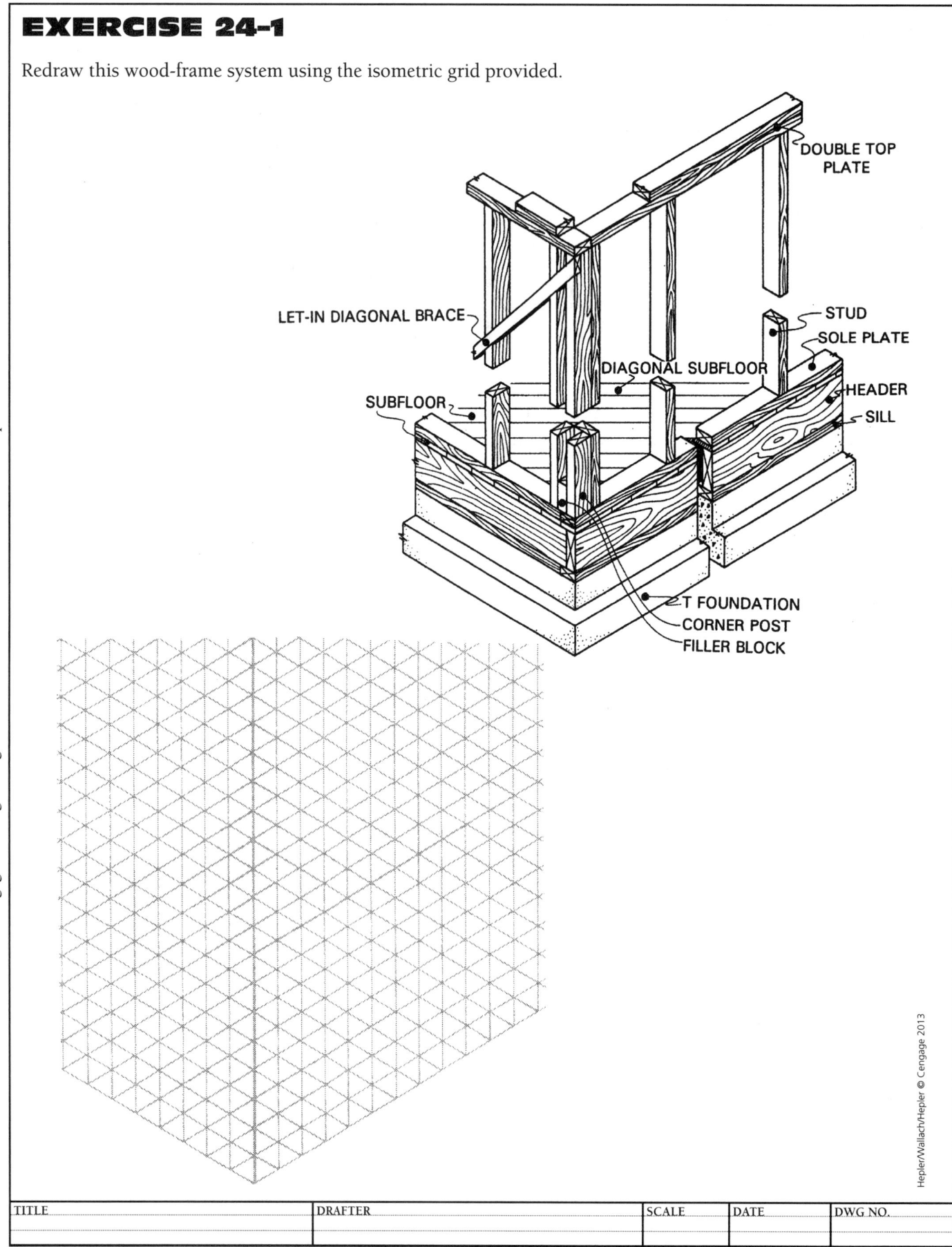

EXERCISE 24 Wood-Frame Systems ■ **127**

EXERCISE 24-2

Using the grid, sketch two wall framing elevations for the post-and-beam construction systems shown.

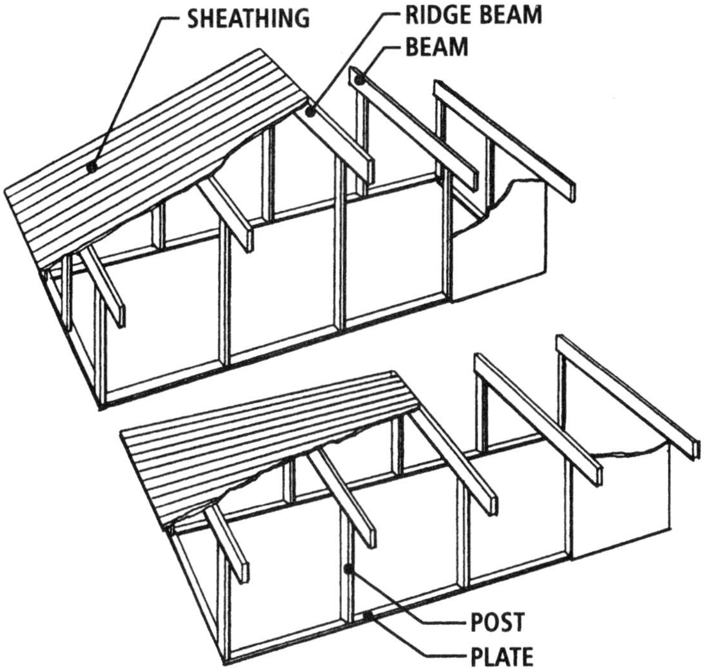

- SHEATHING
- RIDGE BEAM
- BEAM
- POST
- PLATE

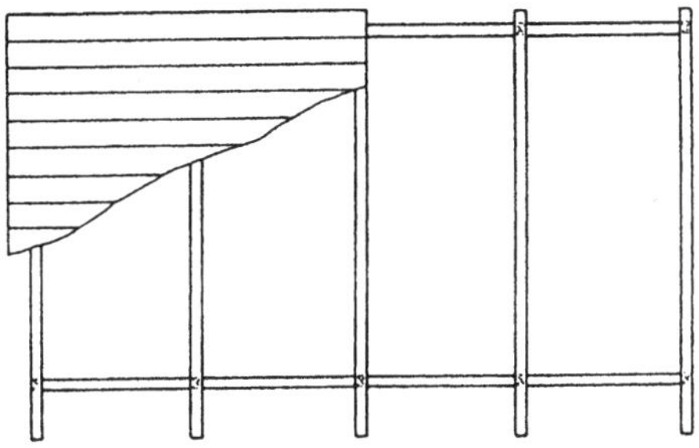

PARTIAL ROOF PLAN

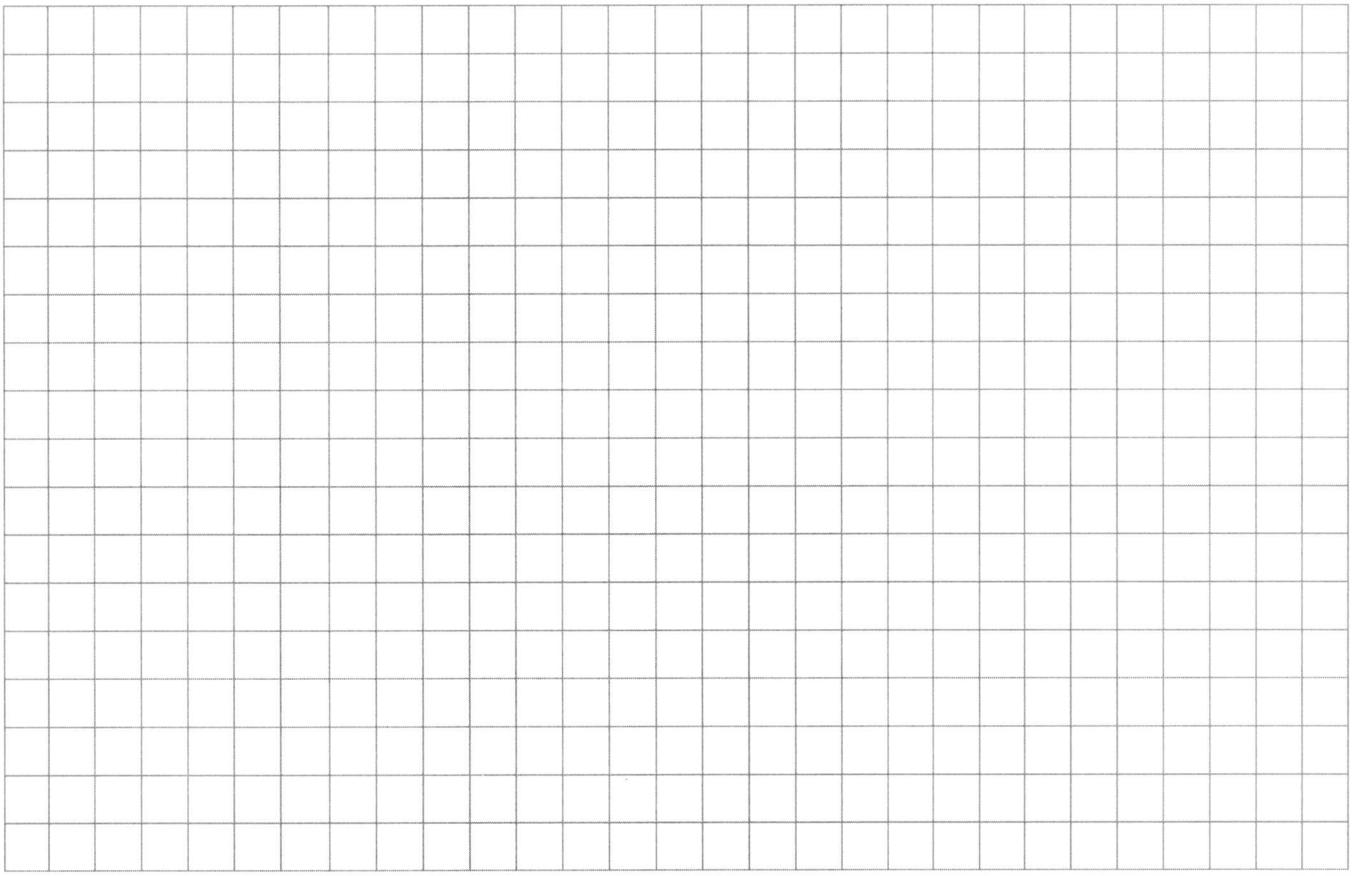

TITLE	DRAFTER	SCALE	DATE	DWG NO.

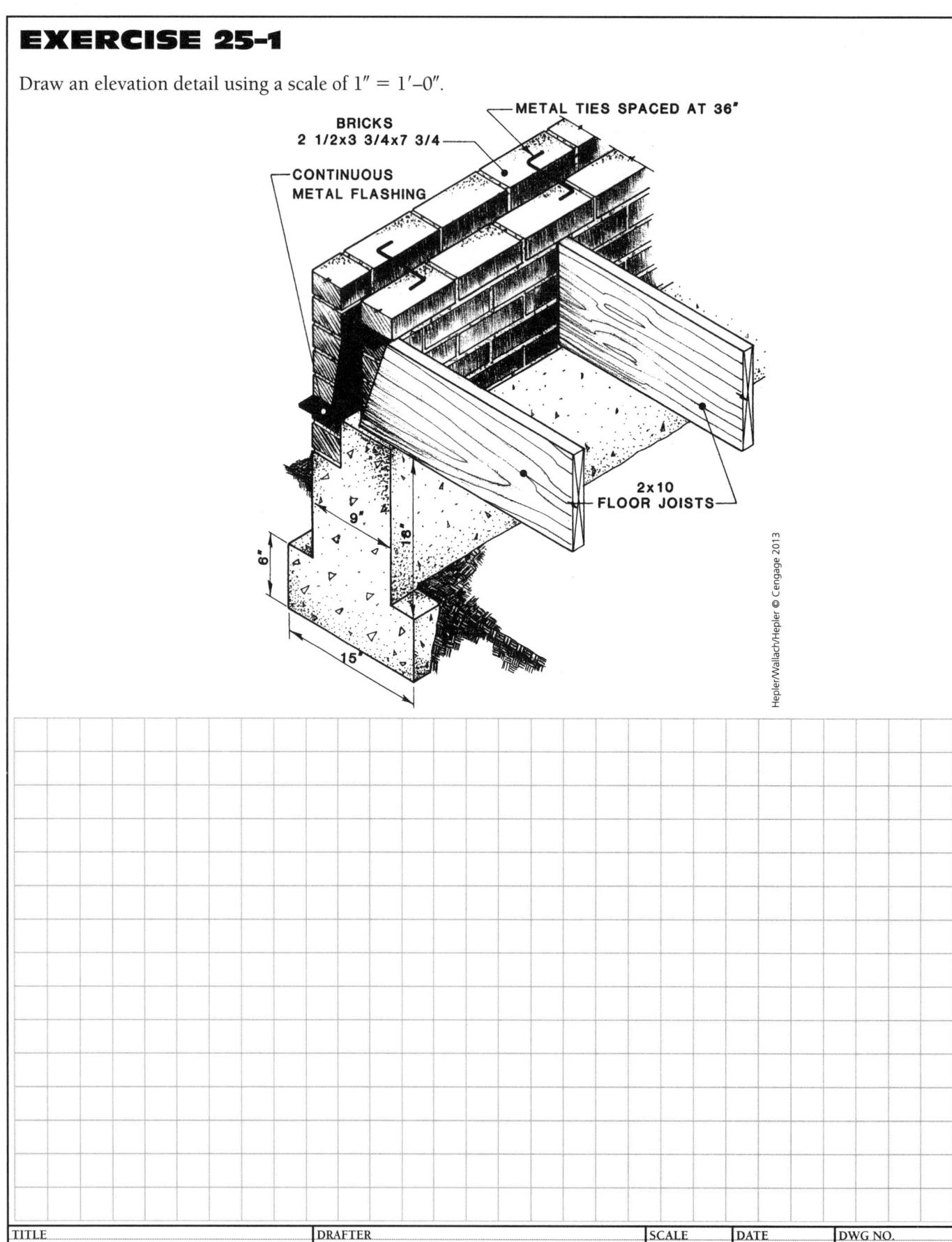

EXERCISE 25-1

Draw an elevation detail using a scale of 1" = 1'–0".

- METAL TIES SPACED AT 36"
- BRICKS 2 1/2 x 3 3/4 x 7 3/4
- CONTINUOUS METAL FLASHING
- 2x10 FLOOR JOISTS
- 9"
- 6"
- 18"
- 15"

TITLE	DRAFTER	SCALE	DATE	DWG NO.

EXERCISE 25-2

Redraw these different types of mortar joints.

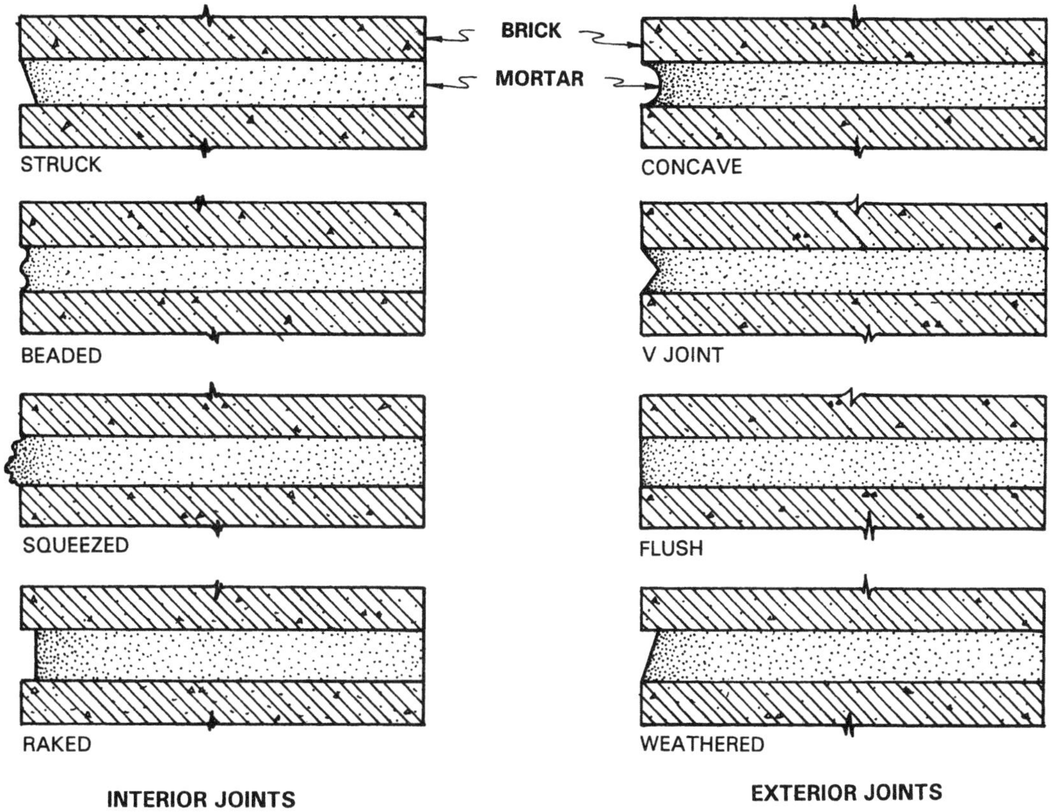
BRICK
MORTAR
STRUCK
CONCAVE
BEADED
V JOINT
SQUEEZED
FLUSH
RAKED
WEATHERED
INTERIOR JOINTS
EXTERIOR JOINTS

TITLE	DRAFTER	SCALE	DATE	DWG NO.

EXERCISE 26-1

Redraw the steel, masonry, and brick construction detail at a scale of 1 1/2" = 1'–0".

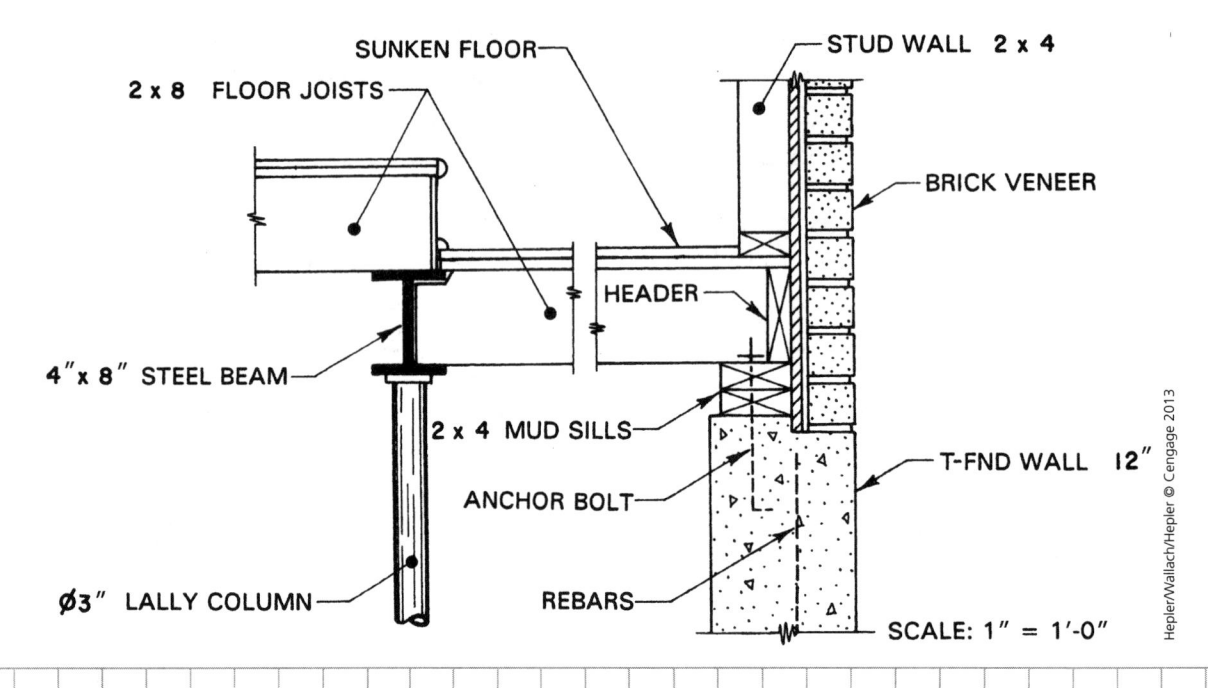

- SUNKEN FLOOR
- 2 x 8 FLOOR JOISTS
- STUD WALL 2 x 4
- BRICK VENEER
- HEADER
- 4" x 8" STEEL BEAM
- 2 x 4 MUD SILLS
- ANCHOR BOLT
- T-FND WALL 12"
- Ø3" LALLY COLUMN
- REBARS
- SCALE: 1" = 1'-0"

TITLE	DRAFTER	SCALE	DATE	DWG NO.

EXERCISE 27 Disaster Prevention Design ■ 131

EXERCISE 27-1

Sketch in steel and wood additions to this floor and T foundation to make them more resistant to hurricanes and earthquakes.

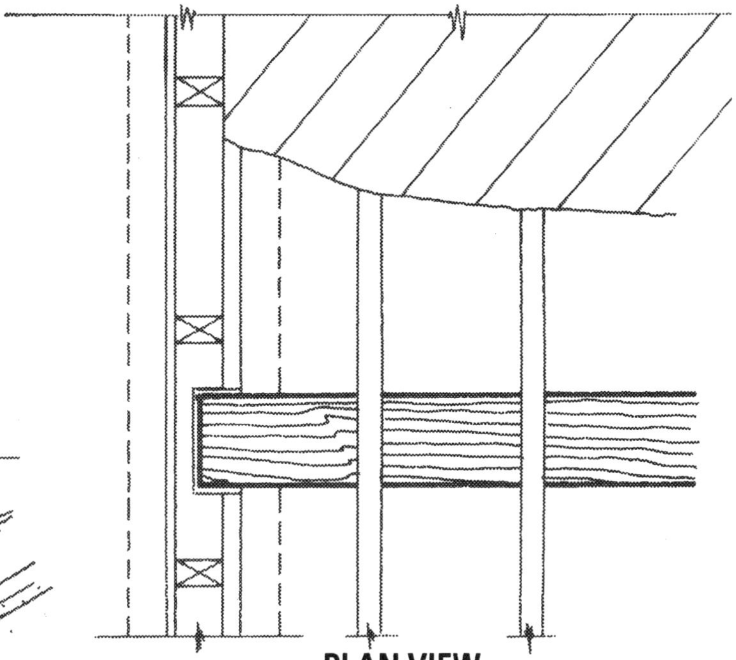

PLAN VIEW

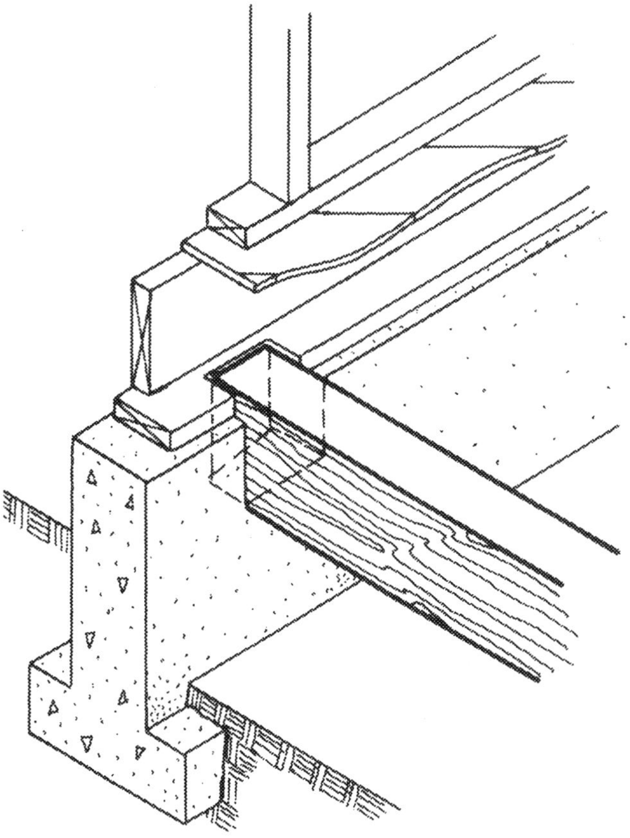

PICTORIAL

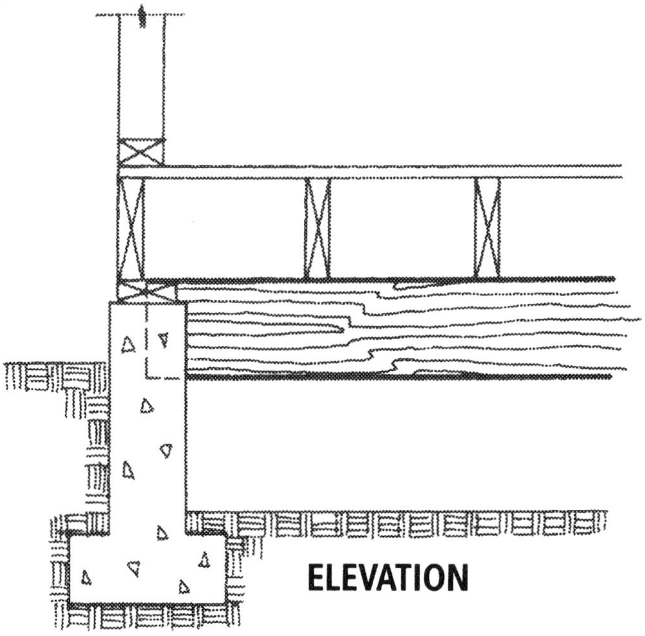

ELEVATION

TITLE	DRAFTER	SCALE	DATE	DWG NO.

EXERCISE 27-2

Below is a T foundation for a one-story residence. Sketch in as many steel and wood additions as necessary to make the floor system more resistant to hurricanes and earthquakes.

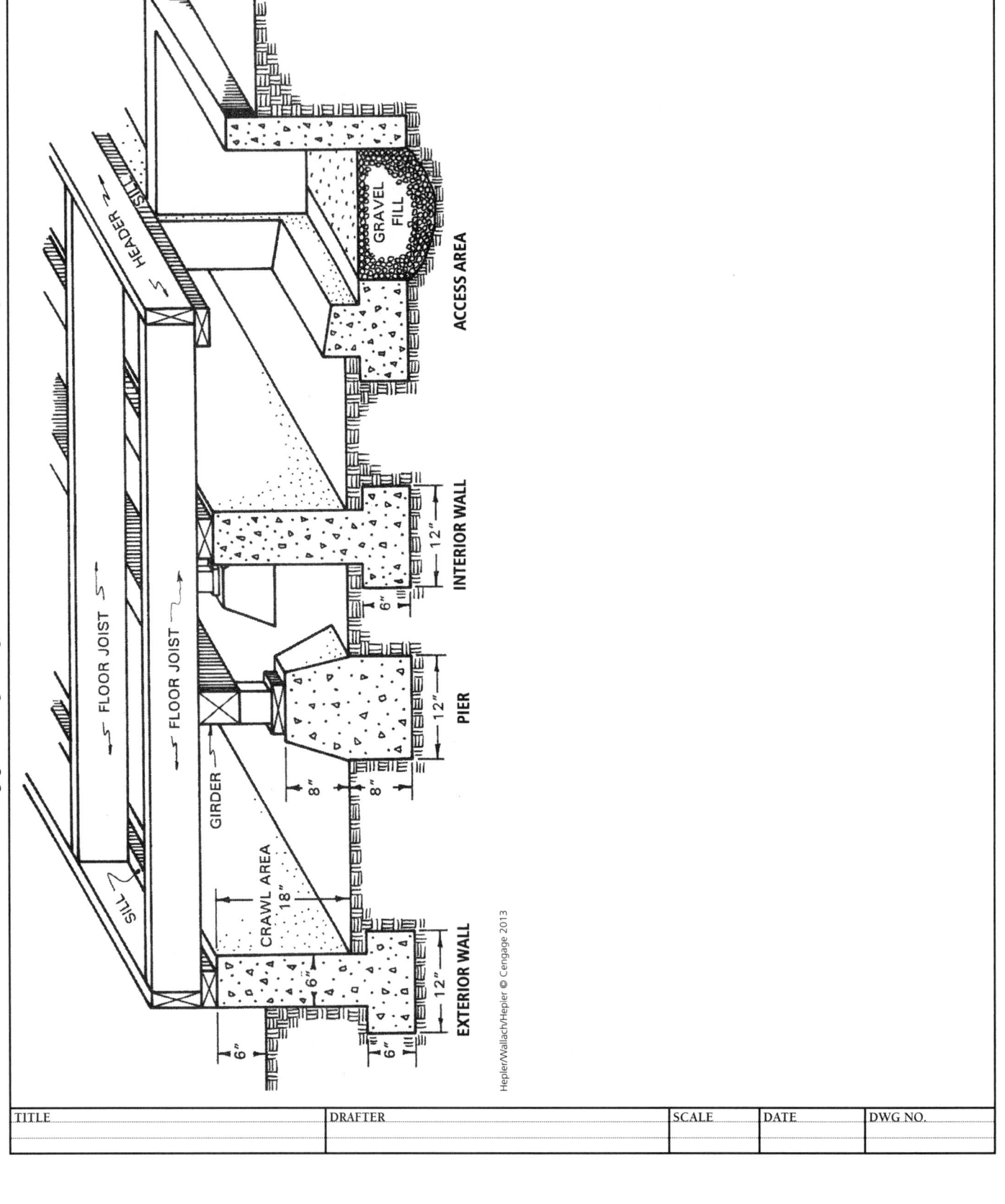

TITLE	DRAFTER	SCALE	DATE	DWG NO.

EXERCISE 28 Floor Framing Drawings ■ 133

EXERCISE 28-1

Study the examples of a floor plan and floor framing plan shown here. Then, referring to the building-code tables in your text, complete the floor framing plan at the bottom of the page. Include the following in your plan:

- Floor joist sizes
- Floor joist spacing
- Girder sizes
- Girder locations
- Pier sizes
- Pier spacing
- Double floor joists under bearing walls parallel to floor joists
- Solid blocking between floor joists under bearing walls perpendicular to floor joists
- Double header framing in all openings in floor system (stairwells, fireplaces)
- Adjacent concrete (garage slab, stairs, patios, walks)

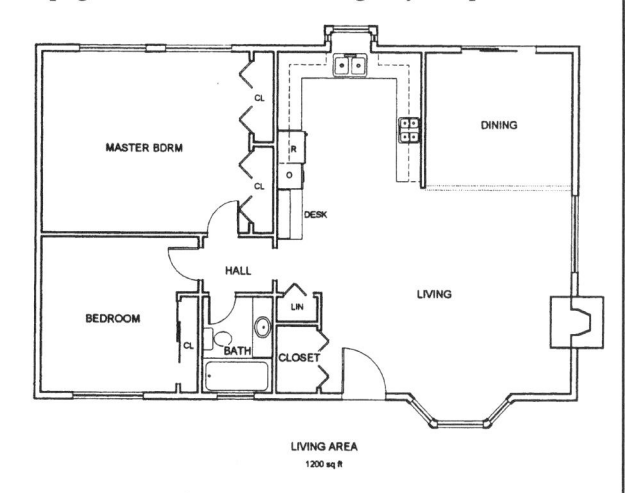

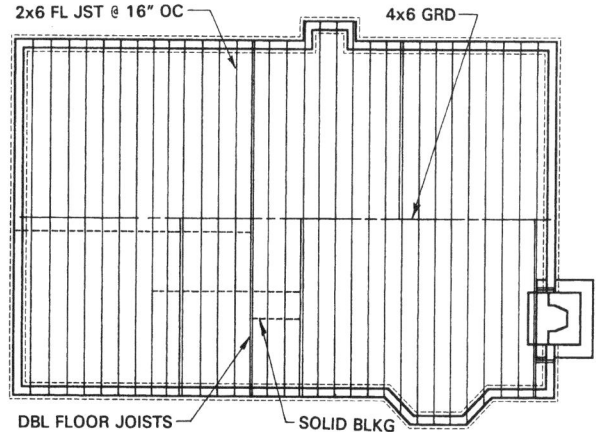

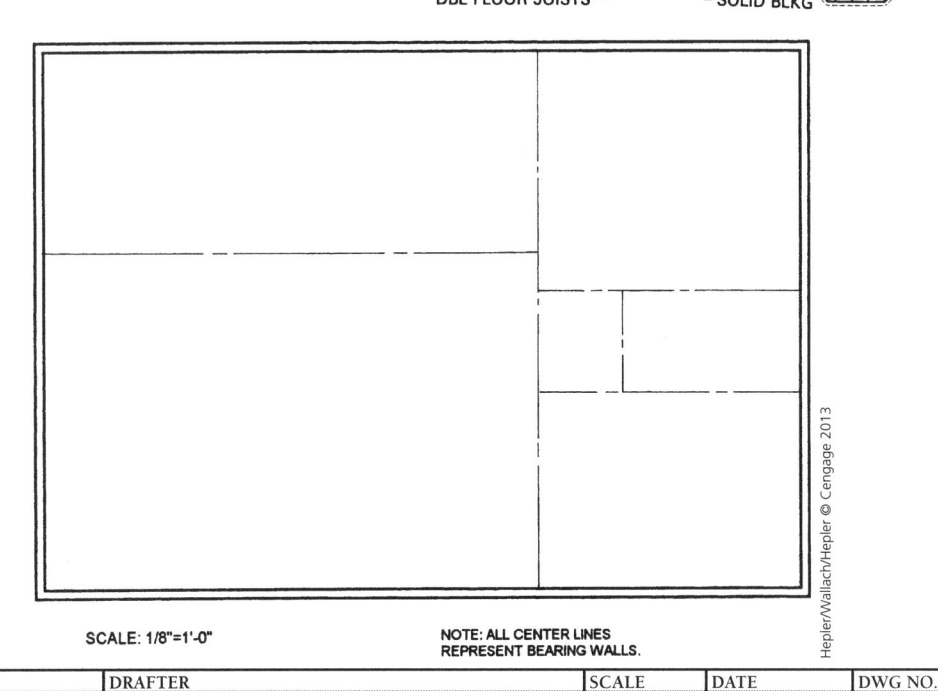

LIVING AREA
1200 sq ft

2x6 FL JST @ 16" OC
4x6 GRD
DBL FLOOR JOISTS
SOLID BLKG

SCALE: 1/8"=1'-0"

NOTE: ALL CENTER LINES REPRESENT BEARING WALLS.

TITLE	DRAFTER	SCALE	DATE	DWG NO.

EXERCISE 28-2

Referring to the pictorial drawings on the left, complete the section views on the right.

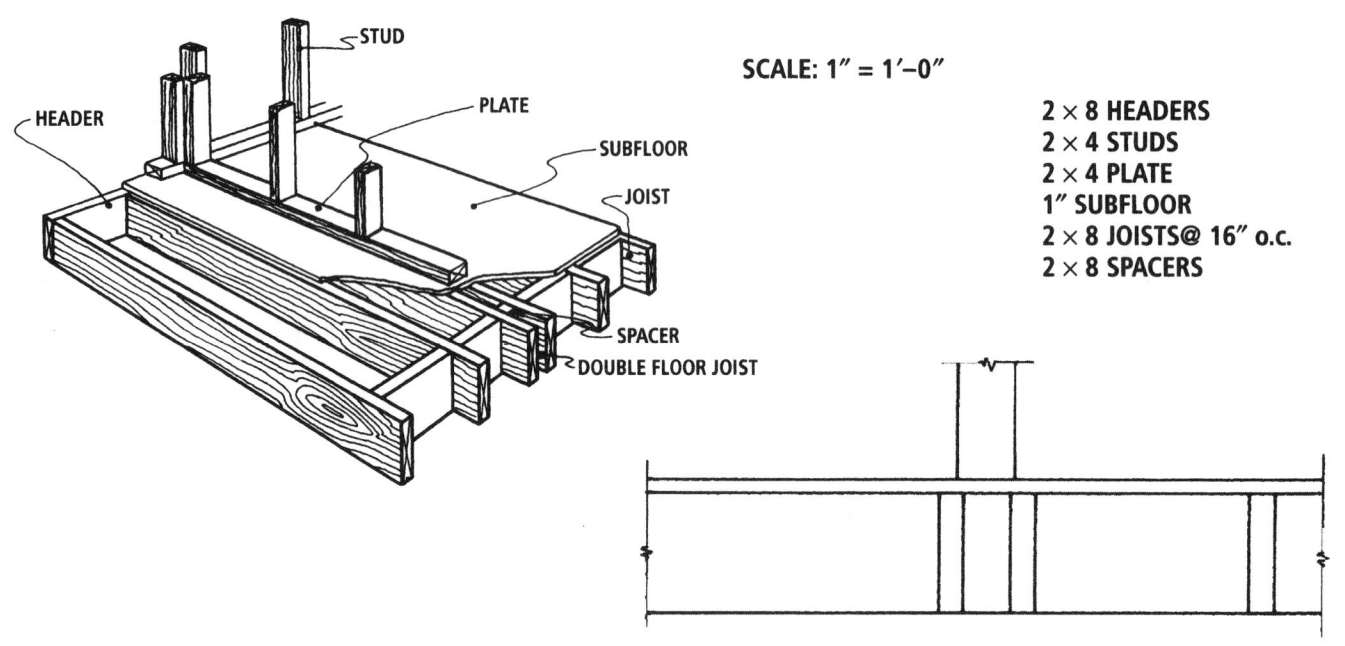

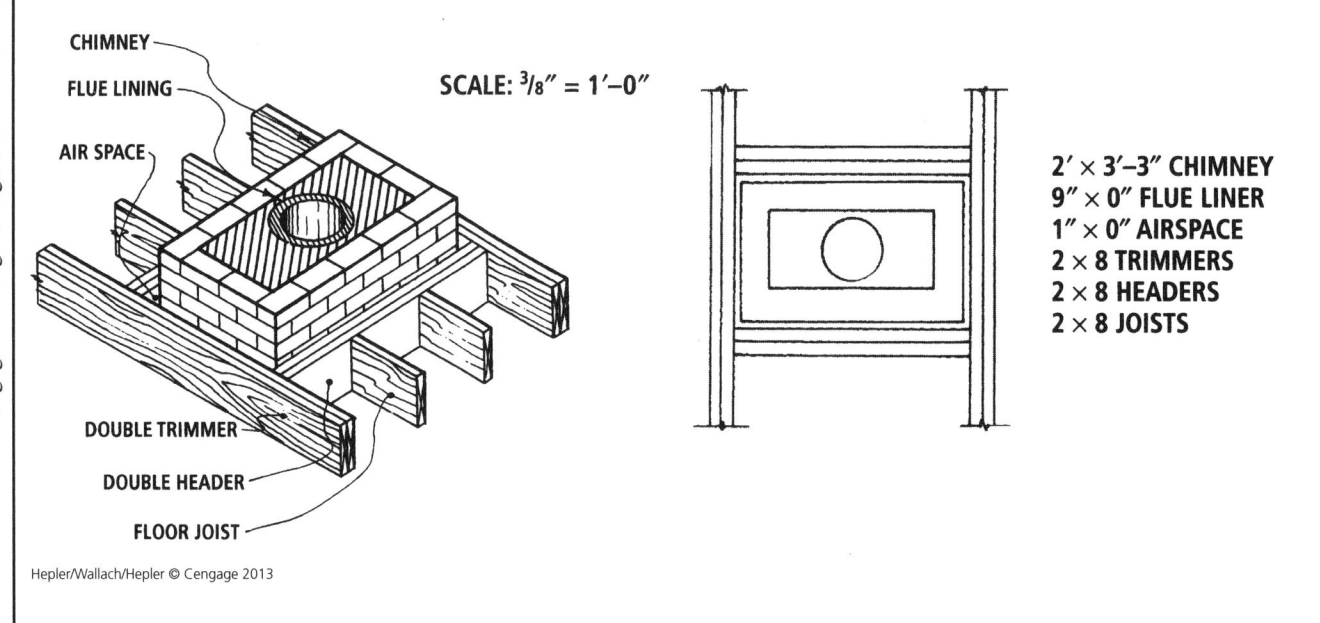

SCALE: 1" = 1'-0"

2 × 8 HEADERS
2 × 4 STUDS
2 × 4 PLATE
1" SUBFLOOR
2 × 8 JOISTS @ 16" o.c.
2 × 8 SPACERS

SCALE: 3/8" = 1'-0"

2' × 3'-3" CHIMNEY
9" × 0" FLUE LINER
1" × 0" AIRSPACE
2 × 8 TRIMMERS
2 × 8 HEADERS
2 × 8 JOISTS

TITLE	DRAFTER	SCALE	DATE	DWG NO.

EXERCISE 28-3

Complete a stud layout on the top plan. Complete a wall framing plan on the bottom elevation.

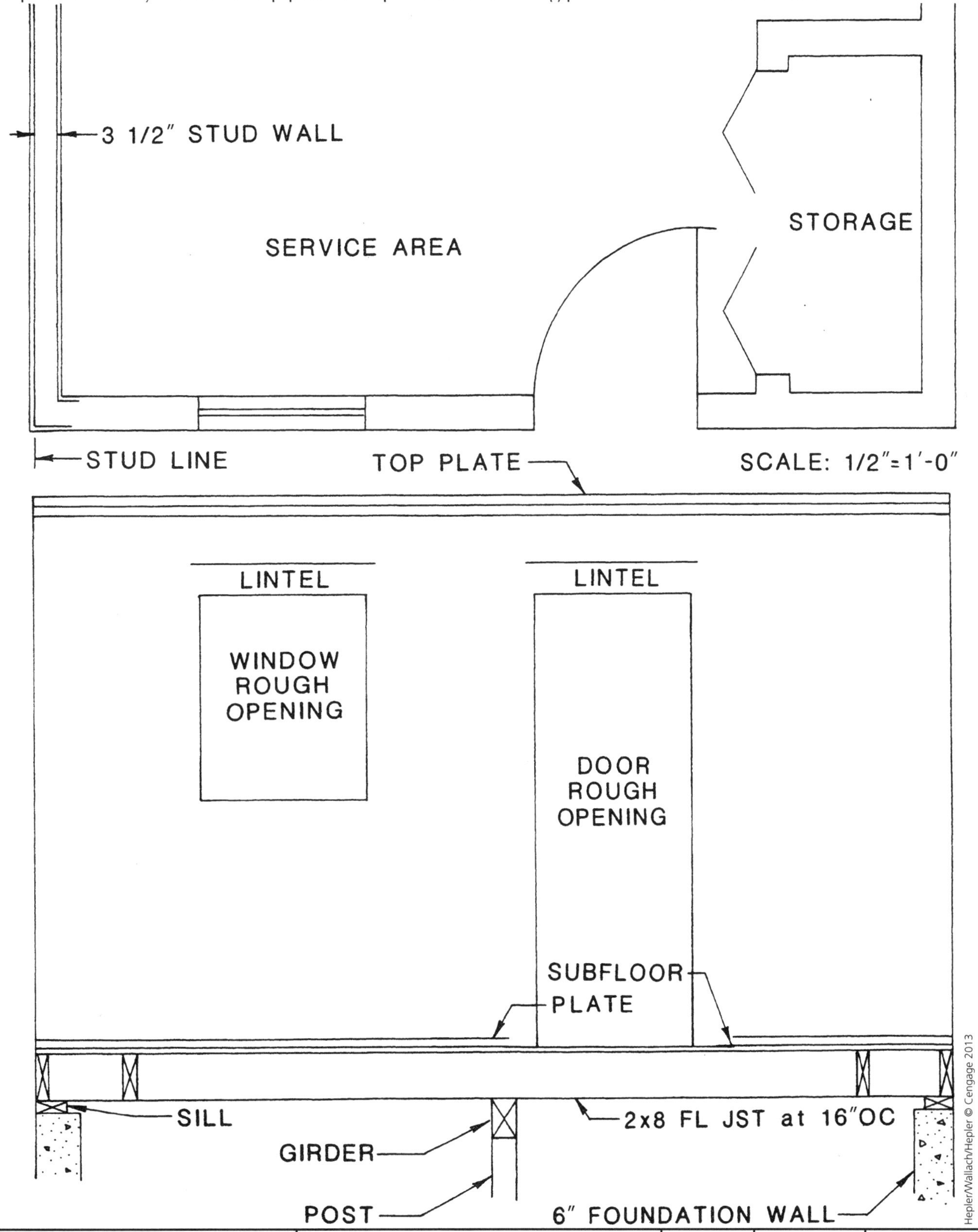
3 1/2″ STUD WALL
SERVICE AREA
STORAGE
STUD LINE
TOP PLATE
SCALE: 1/2″=1′-0″
LINTEL
LINTEL
WINDOW ROUGH OPENING
DOOR ROUGH OPENING
SUBFLOOR
PLATE
SILL
GIRDER
2x8 FL JST at 16″OC
POST
6″ FOUNDATION WALL
Hepler/Wallach/Hepler © Cengage 2013

TITLE	DRAFTER	SCALE	DATE	DWG NO.

EXERCISE 28-4

Draw a wall framing plan below using the stud layout plan as a guide.

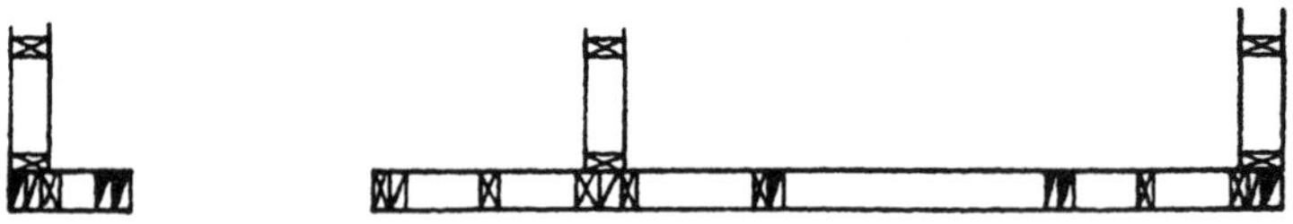

DATA FOR WALL FRAMING AND STUD LAYOUT PLANS

1. Size of top plate
2. Type of corner framing
3. Type of interior wall intersection
4. 2x4 studs spaced at 16° OC
5. Lintel sizes (check building codes)
6. Plate sizes
7. Subfloor size
8. Spacers
9. Cripple studs
10. Types of diagonal bracing

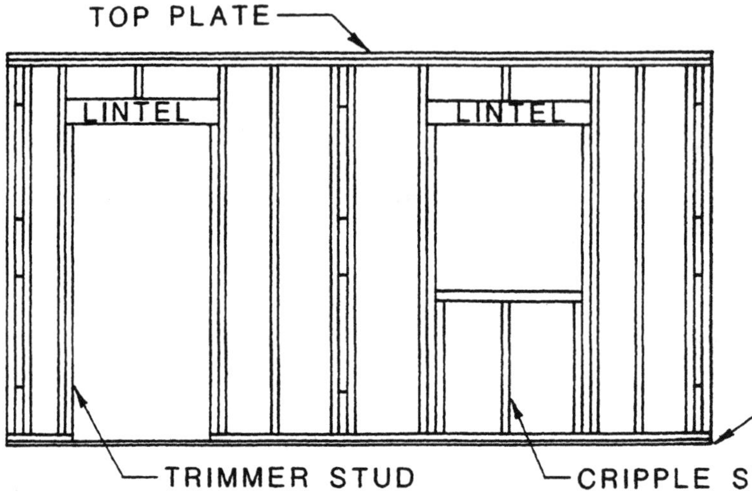

SCALE: 1/4"=1'-0"

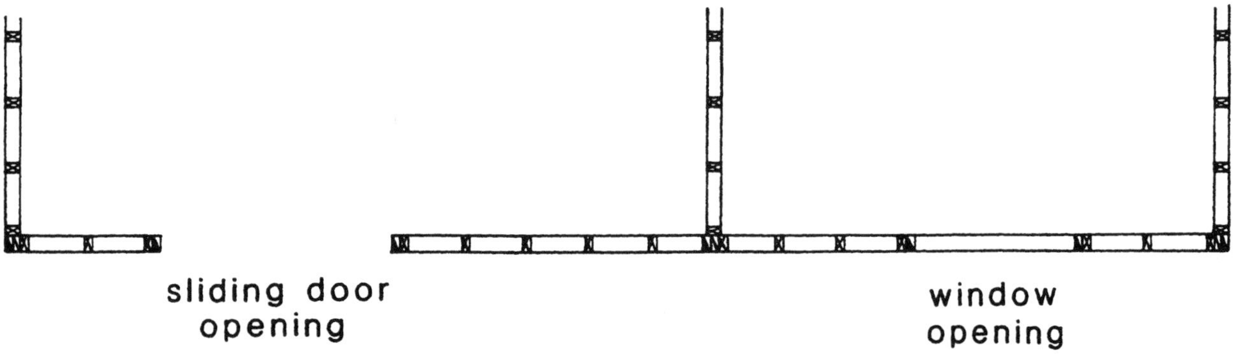

sliding door opening

window opening

SCALE: 1/4"=1'-0"

TITLE	DRAFTER	SCALE	DATE	DWG NO.

EXERCISE 28-5

Complete the stud layout plan below.

SCALE: 1/2" = 1'-0"

WINDOW FRAMING

DOOR FRAMING

STUD

CORNER FRAMING

INTERIOR WALL INTERSECTION FRAMING

TRIMMER STUD

STUD LAYOUT PLAN DATA
1. Finished stud size 1 1/2" x 3 1/2"
2. Studs spaced at 16" OC ⊠
3. Studs not on 16" modules ◣
4. Spacer studs ◻
5. Doors and windows framed with studs, trimmers, lintels, sills

SCALE: 1/2" = 1'-0"

3 1/2" STUD WALL

EXTERIOR WALL LINE
INTERIOR WALL LINE

| TITLE | DRAFTER | SCALE | DATE | DWG NO. |

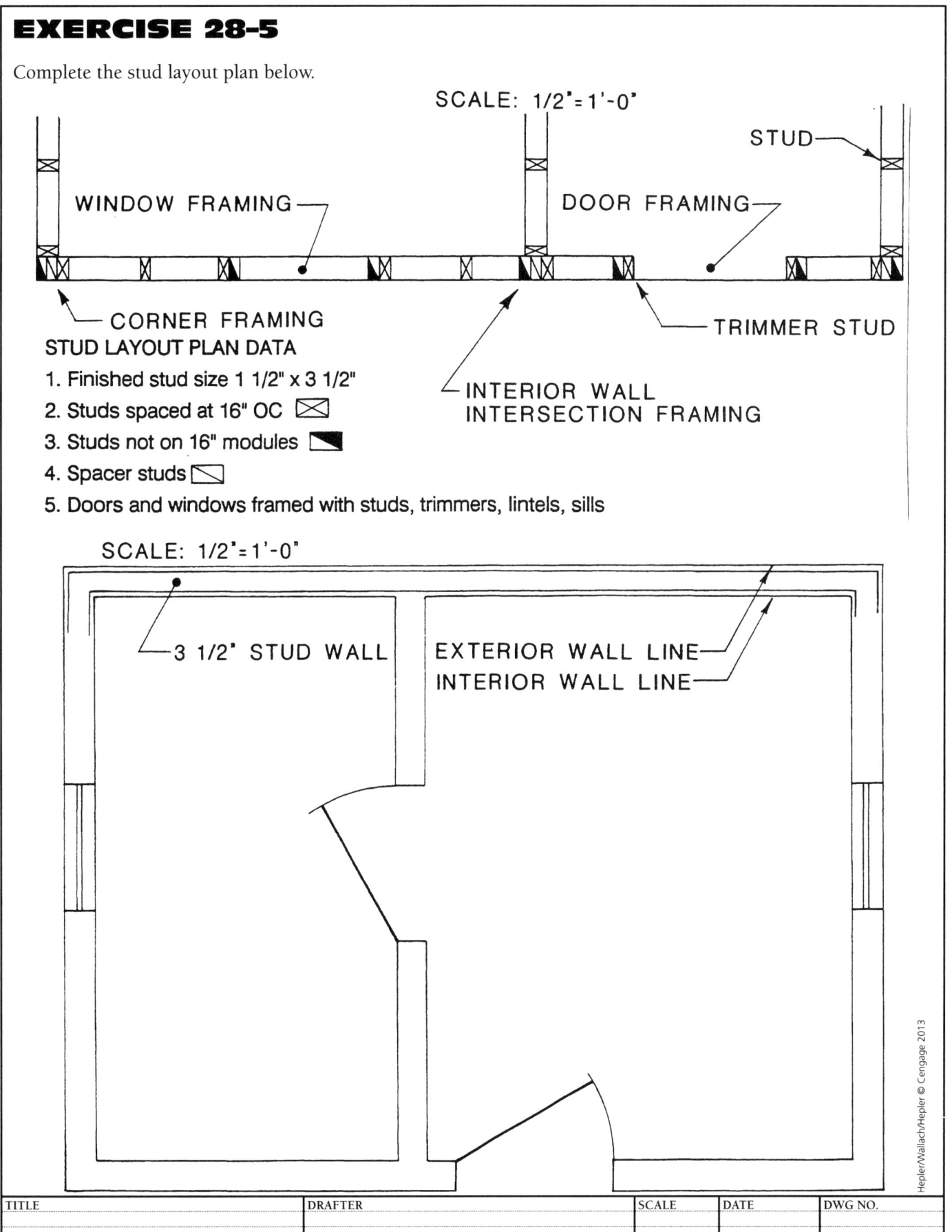

EXERCISE 28-6

Complete the stud layout plan using the bottom elevation as a guide.

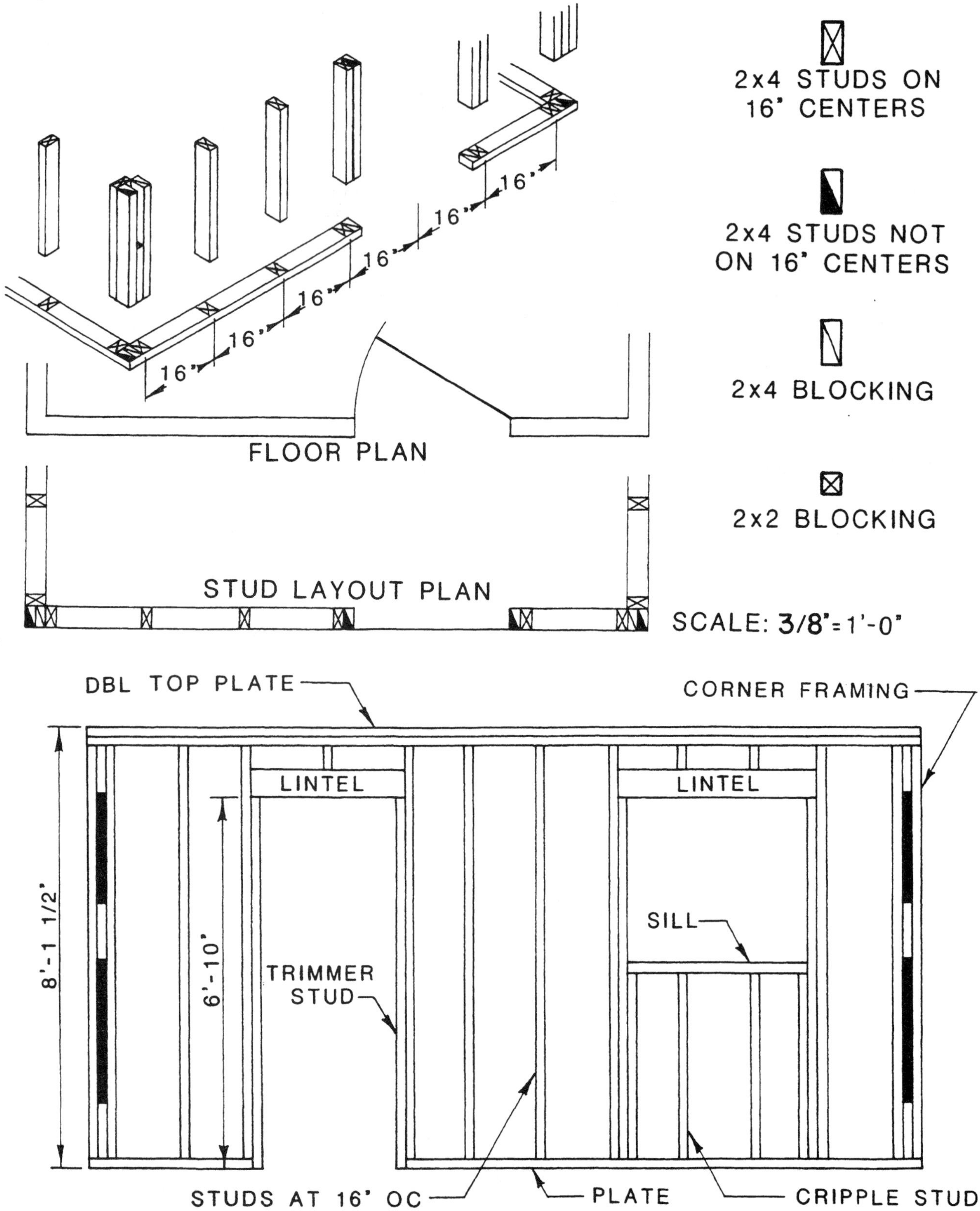

16"
16"
16"
16"
16"
16"
2x4 STUDS ON 16" CENTERS
2x4 STUDS NOT ON 16" CENTERS
2x4 BLOCKING
2x2 BLOCKING
FLOOR PLAN
STUD LAYOUT PLAN
SCALE: 3/8"=1'-0"
DBL TOP PLATE
CORNER FRAMING
LINTEL
LINTEL
8'-1 1/2"
6'-10"
SILL
TRIMMER STUD
STUDS AT 16" OC
PLATE
CRIPPLE STUD

TITLE	DRAFTER	SCALE	DATE	DWG NO.

EXERCISE 28-7

Complete the floor framing plan. Center lines represent interior bearing walls. Scale: 1/4″ = 1′–0″.

TITLE	DRAFTER	SCALE	DATE	DWG NO.

EXERCISE 29-1

Study the floor plan and the pictorial stud layout for the closet. Then complete the stud layout drawing at the bottom of the page.

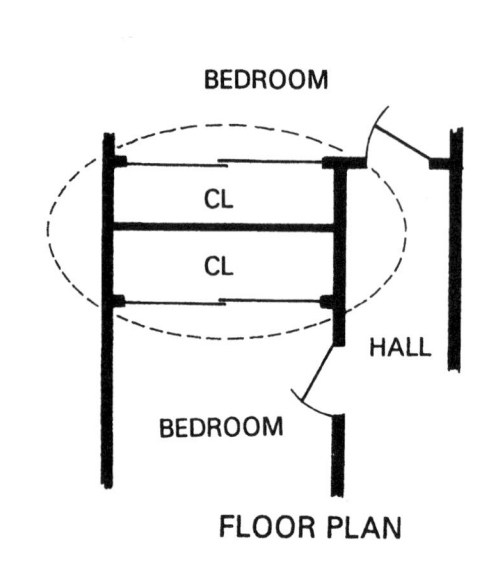

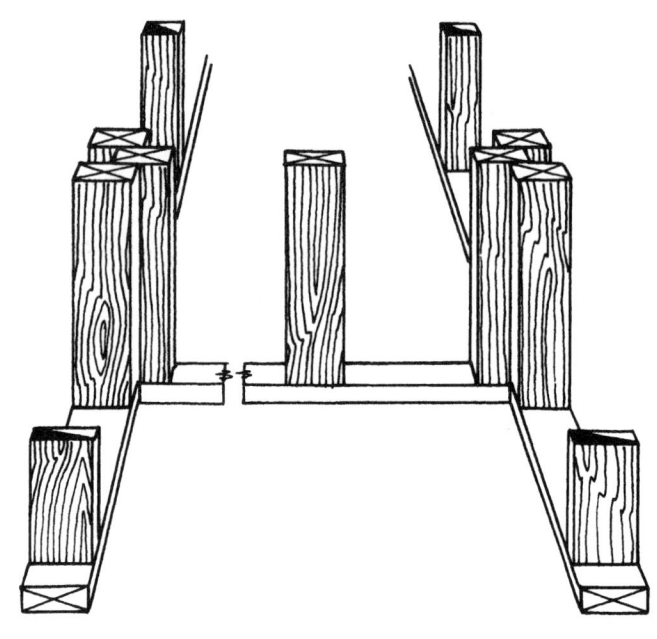

FLOOR PLAN

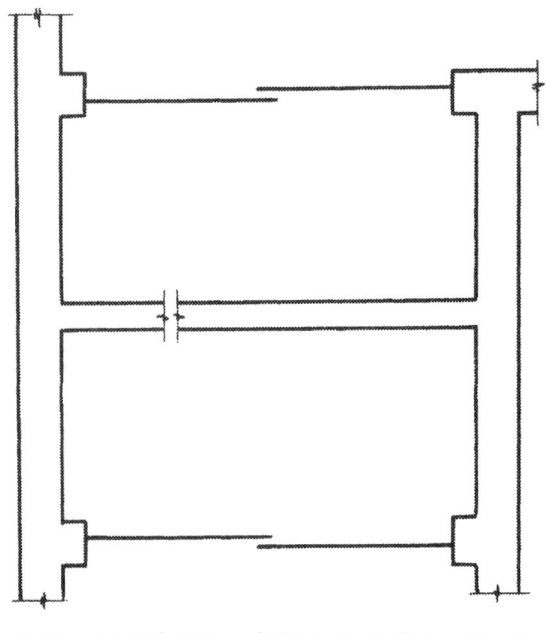

PICTORIAL STUD LAYOUT

STUD LAYOUT   SCALE: 1/2" = 1'-0"

TITLE	DRAFTER	SCALE	DATE	DWG NO.

EXERCISE 29-2

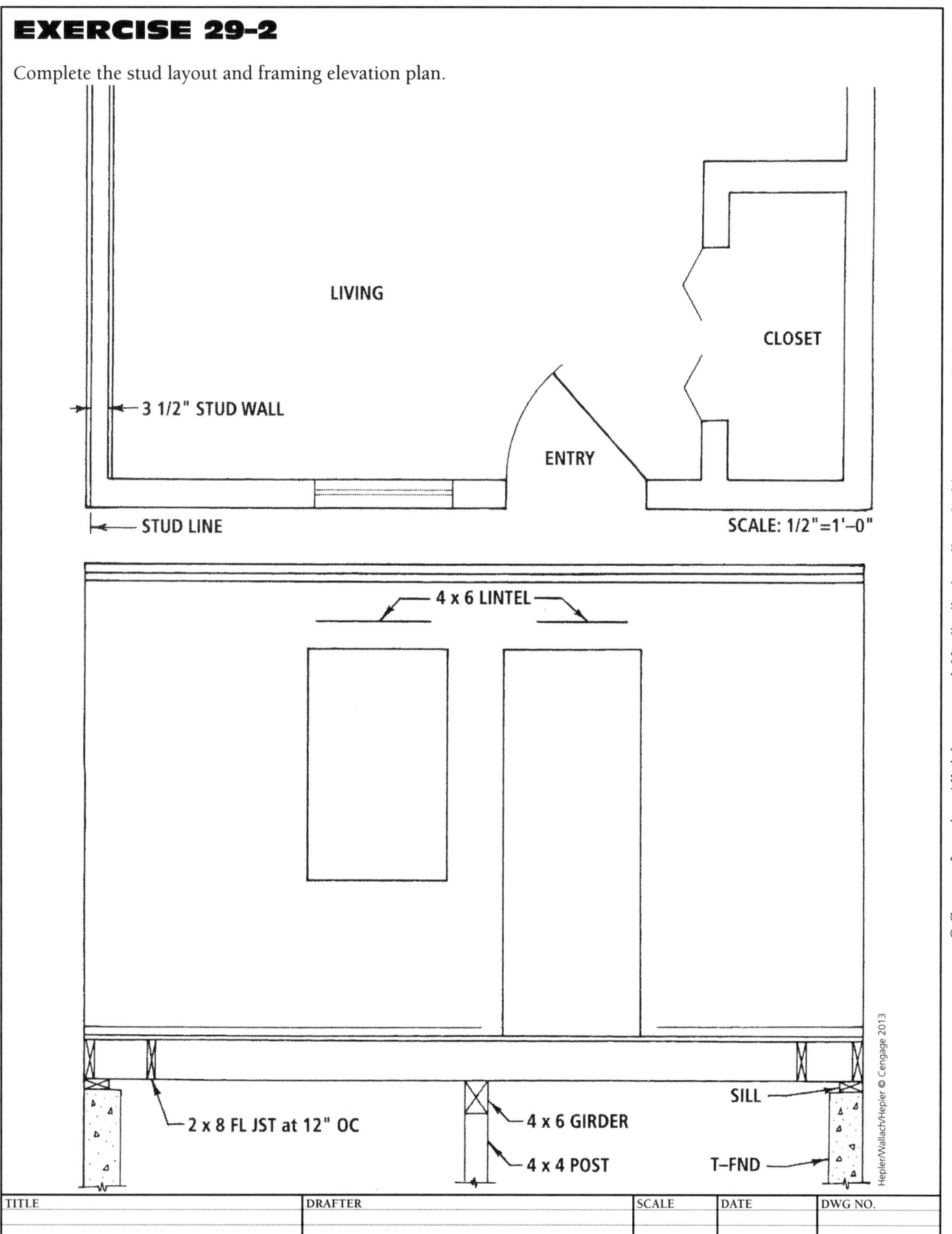

Complete the stud layout and framing elevation plan.

3 1/2" STUD WALL

LIVING

CLOSET

ENTRY

STUD LINE

SCALE: 1/2"=1'-0"

4 x 6 LINTEL

2 x 8 FL JST at 12" OC

4 x 6 GIRDER

4 x 4 POST

SILL

T-FND

TITLE	DRAFTER	SCALE	DATE	DWG NO.

EXERCISE 30-1

Using the grid, draw a single-line roof framing plan for the hip roof shown. Use 2 × 8 rafters 24" OC.

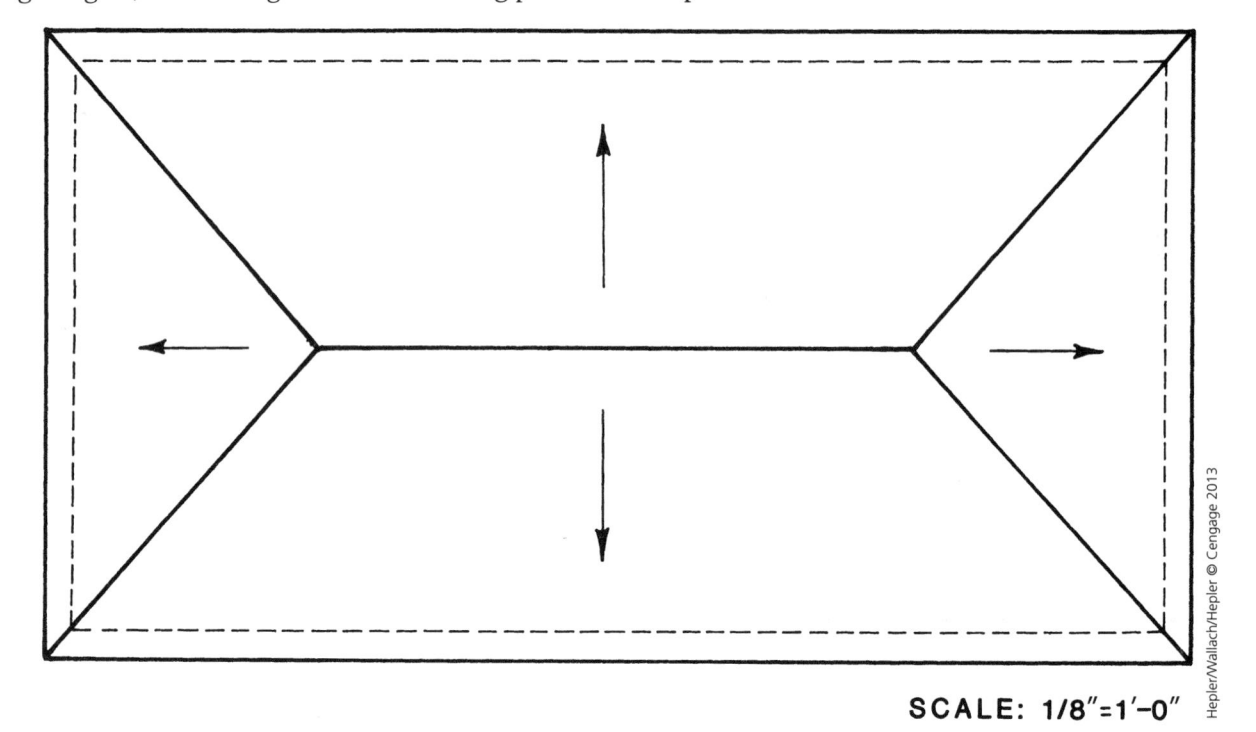

SCALE: 1/8"=1'-0"

TITLE	DRAFTER	SCALE	DATE	DWG NO.

EXERCISE 30 Roof Framing Drawings ■ 143

EXERCISE 30-2, PART A

Draw a gable roof framing plan for this cabin. Use 2 × 6 rafters at 16″ OC, a 2 × 8 ridge board, and a 6″ overhang on all sides. Use any convenient roof pitch. Scale: 1/4″ = 1′–0″.

EXERCISE 30-2, PART B

Draw a single-line hip roof framing plan for this cabin. Use 2 × 8 rafters at 16″ OC, a 2 × 10 ridge board, a 12″ overhang on all sides, and a 45° hip angle. Scale: 1/4″ = 1′–0″.

TITLE	DRAFTER	SCALE	DATE	DWG NO.

EXERCISE 30-3

Using the pictorials as a guide, complete the cornice sections. Roof pitch is 30°, and overhang is 6″. Use 2 × 6 rafters, 2 × 4 studs, 1″ sheathing, 1″ bevel siding, 1″ airspace, 3″ brick veneer, 6″ solid brick, and a 2 × 8 top plate. Scale: 1″ = 1′–0″.

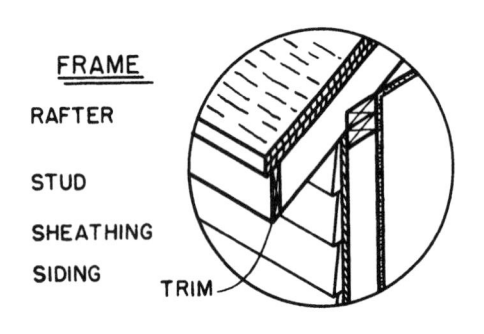

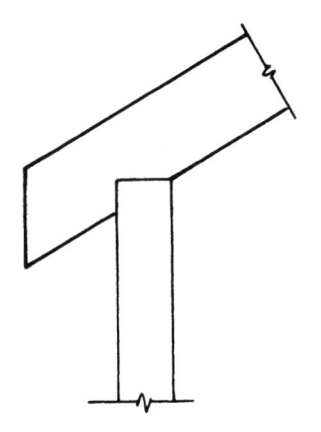

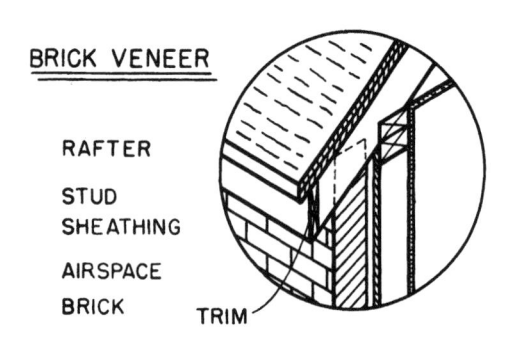

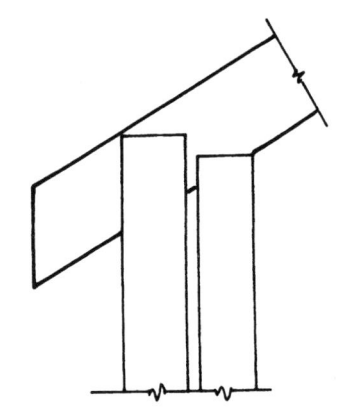

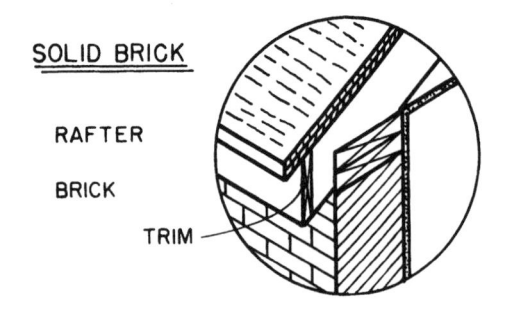

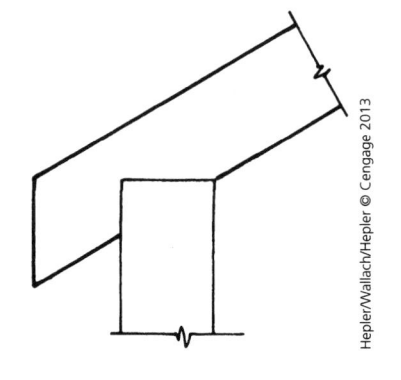

FRAME
- RAFTER
- STUD
- SHEATHING
- SIDING
- TRIM

BRICK VENEER
- RAFTER
- STUD
- SHEATHING
- AIRSPACE
- BRICK
- TRIM

SOLID BRICK
- RAFTER
- BRICK
- TRIM

TITLE	DRAFTER	SCALE	DATE	DWG NO.

EXERCISE 30-4, PART A

Referring to the pictorial, complete the dormer framing plan and elevation.

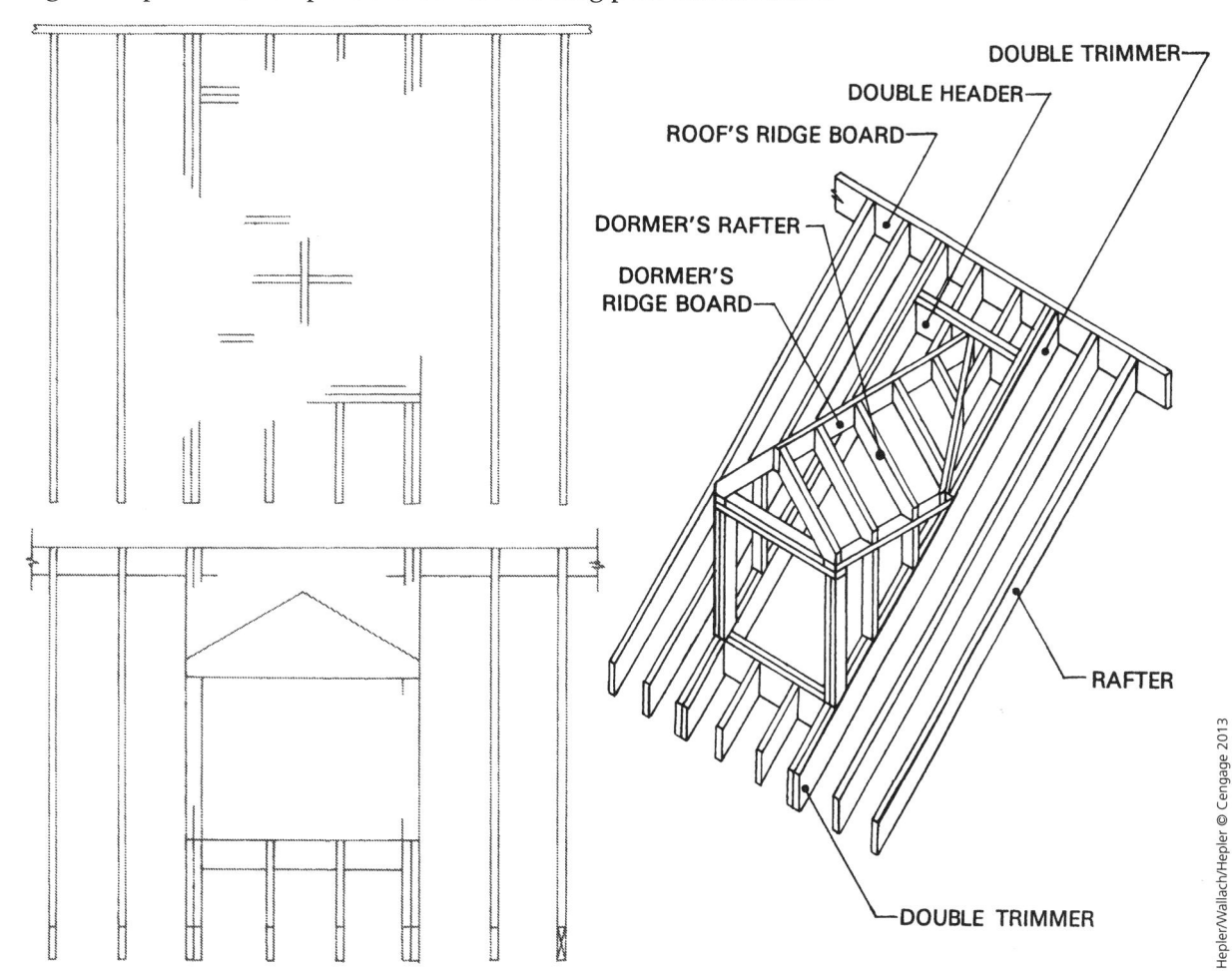

DOUBLE TRIMMER
DOUBLE HEADER
ROOF'S RIDGE BOARD
DORMER'S RAFTER
DORMER'S RIDGE BOARD
RAFTER
DOUBLE TRIMMER

EXERCISE 30-4, PART B

Use the roof plan to complete a roof framing plan. Refer to the building-code tables in the student text as needed.
Scale: 3/32" = 1'–0"

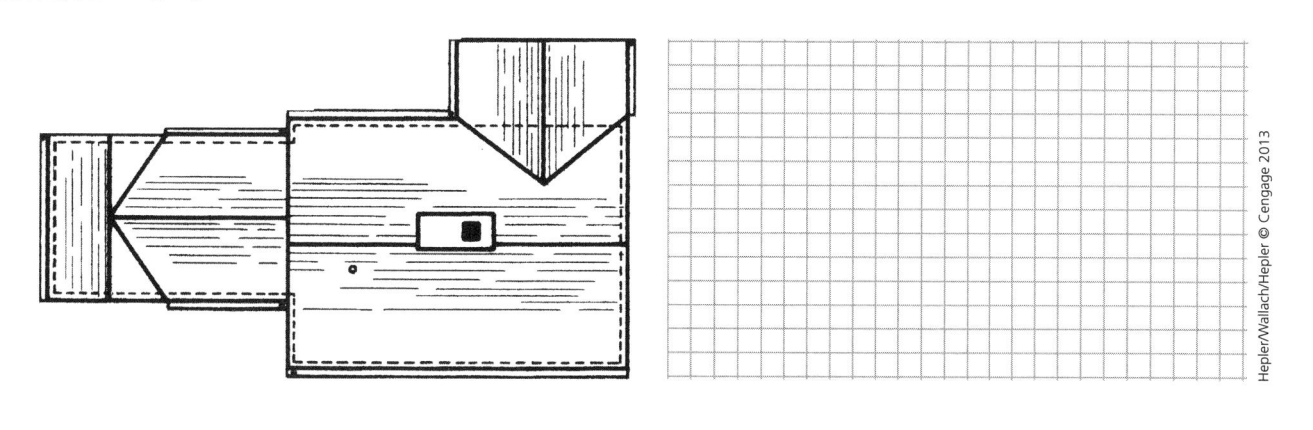

TITLE	DRAFTER	SCALE	DATE	DWG NO.

EXERCISE 30-5

Complete a gable roof framing plan with gable-end lookouts. Use 2″ × 6″ rafters at 16″ OC with a 12″ overhang and a 2″ × 8″ ridge board.

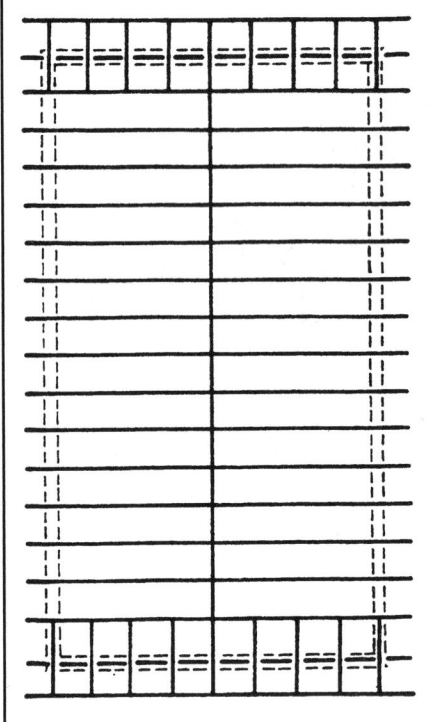

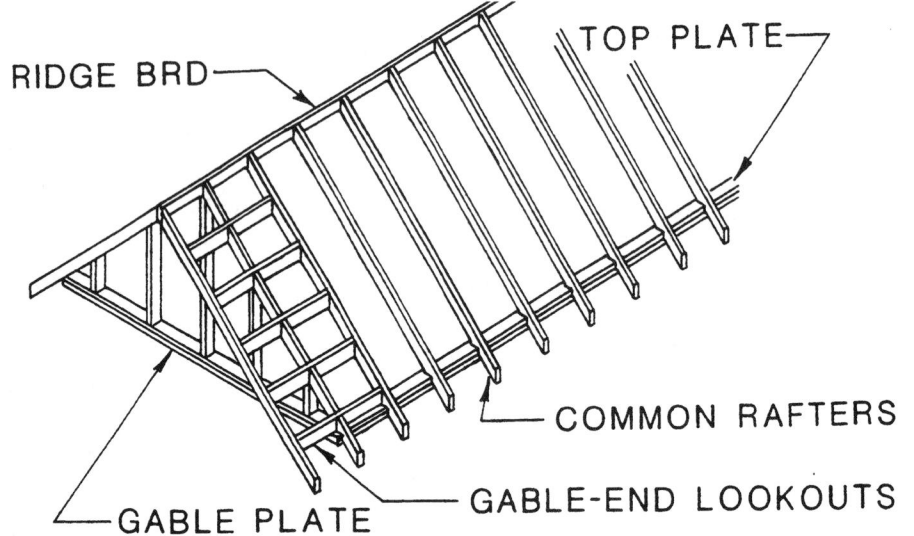

GABLE ROOF FRAMING DATA:
2x6 RF RFTS AT 16"OC
12" CONT OH
2x8 RIDGE BOARD

TITLE	DRAFTER	SCALE	DATE	DWG NO.

EXERCISE 30-6

Complete a single-line hip roof framing plan below with 2″ × 6″ rafters 16″ OC. Show a 12″ overhang, a 2″ × 8″ ridge board, and a hip angle of 45°.

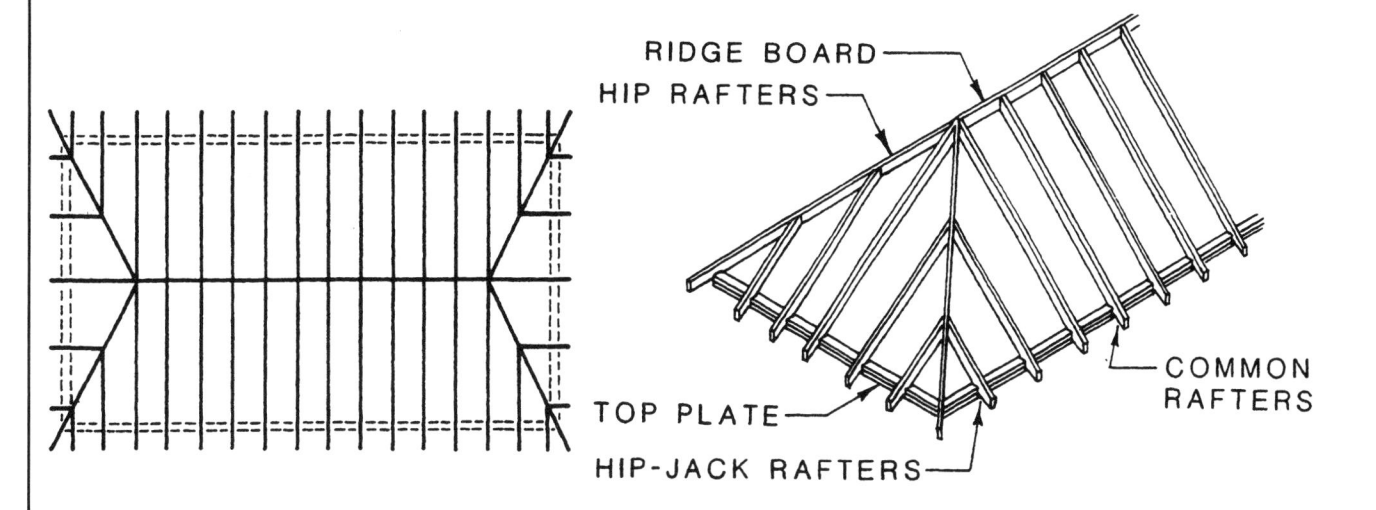

EXERCISE 30-7

Complete a single-line gable roof framing plan with 2″ × 8″ rafters 24″ OC.

SCALE: 1/8″ = 1′-0″

TITLE	DRAFTER	SCALE	DATE	DWG NO.

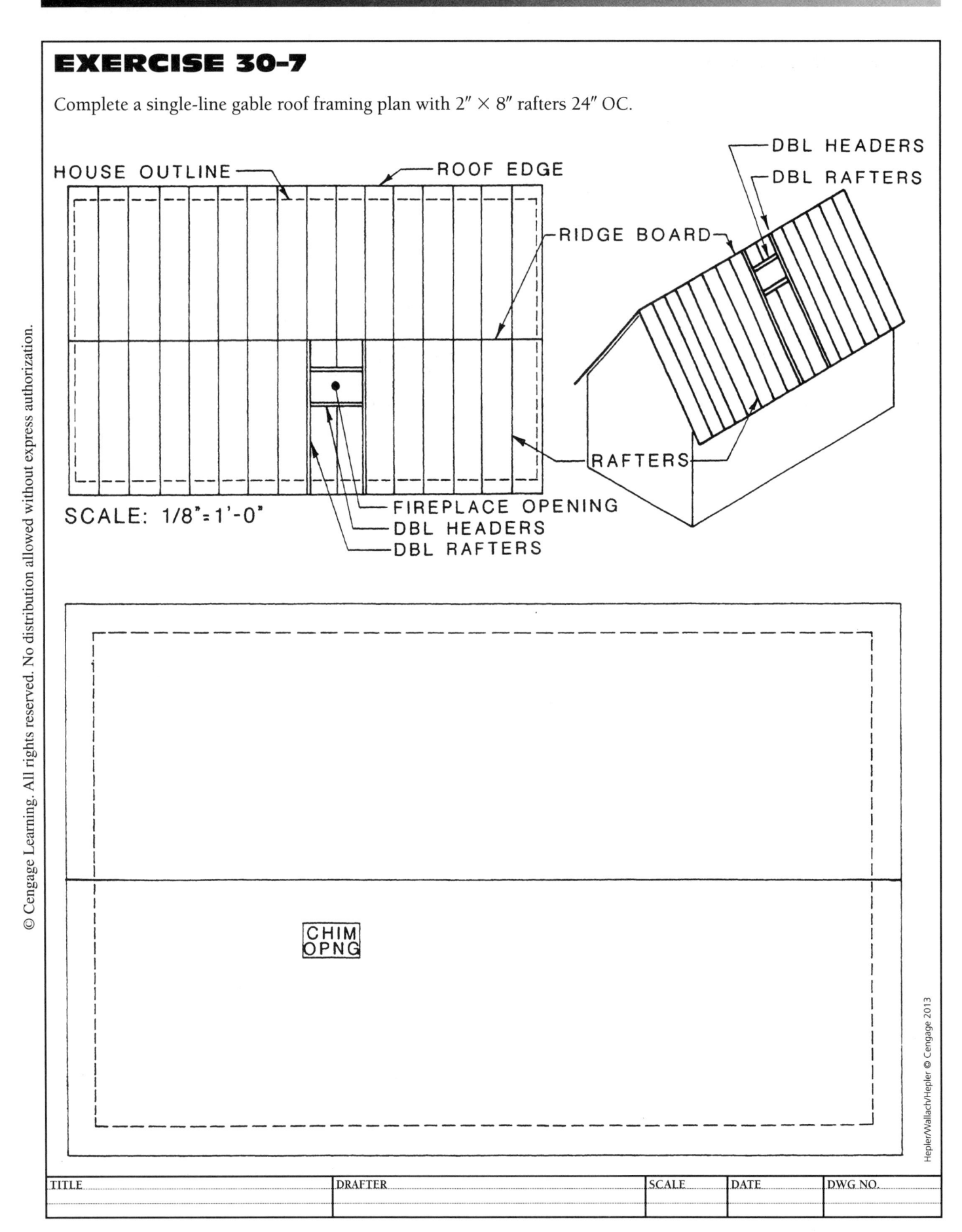

EXERCISE 31 Electrical Design and Drawings ■ **149**

EXERCISE 31-1

Using the appropriate symbols and line conventions, complete the electrical plan for this apartment.

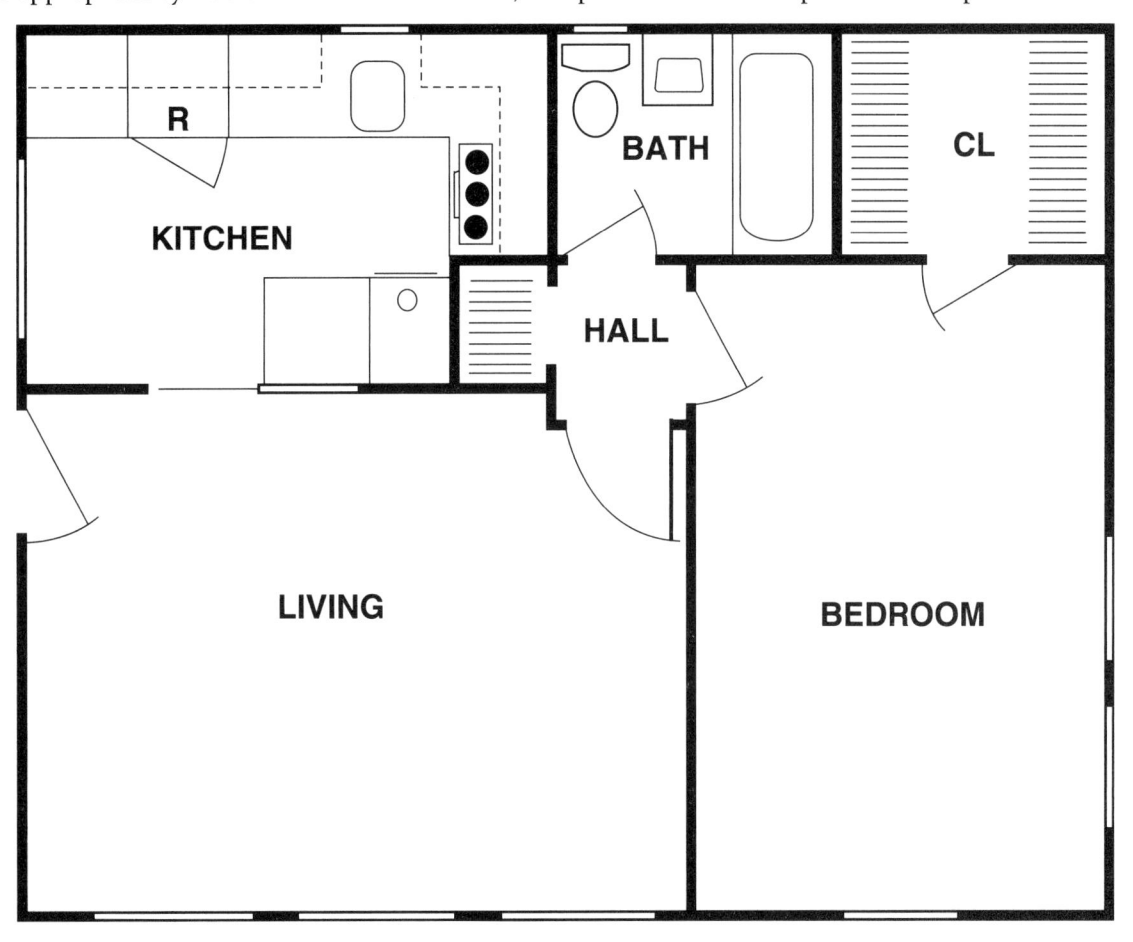

Symbol	Description	Symbol	Description
○	CEILING FIXTURE	⊢○⊣	120 V DOUBLE OUTLET
○○ PS	PULL CORD SWITCH	⊢◎⊣	240 V OUTLET
⊢○	WALL FIXTURE	⊢○⊣	120 V TRIPLE OUTLET
⊢ⓒ	CLOCK	⊢◉ DRY	SPECIAL PURPOSE OUTLET (240 V DRYER)
▨	POWER PANEL	⊢○ WP	WEATHERPROOF OUTLET
◁	TELEPHONE OUTLET	⊢○ GFCI	GROUND-FAULT CIRCUIT INTERRUPTER
◀	TELEPHONE JACK	•	FLOOR OUTLET
▭	BELL	S	SWITCH (TWO WIRES)
Ⓜ	METER	S_3	SWITCH (THREE WIRES)
⊢ G	GAS OUTLET	S_4	SWITCH (FOUR WIRES)
⊠	HEAT OUTLET	S_D	AUTOMATIC DOOR SWITCH
▭ FD	FLOOR DRAIN	S_P	SWITCH WITH PILOT LIGHT
ⓌⒽ	WATER HEATER	S̲	LOW-VOLTAGE SWITCH

TITLE	DRAFTER	SCALE	DATE	DWG NO.

EXERCISE 31-2

Using the appropriate symbols and line conventions, complete the electrical plan for this house.

| TITLE | DRAFTER | SCALE | DATE | DWG NO. |

EXERCISE 31 Electrical Design and Drawings ■ **151**

EXERCISE 31-3

Using the appropriate symbols and line conventions, complete both an indoor and outdoor electrical plan for this house.

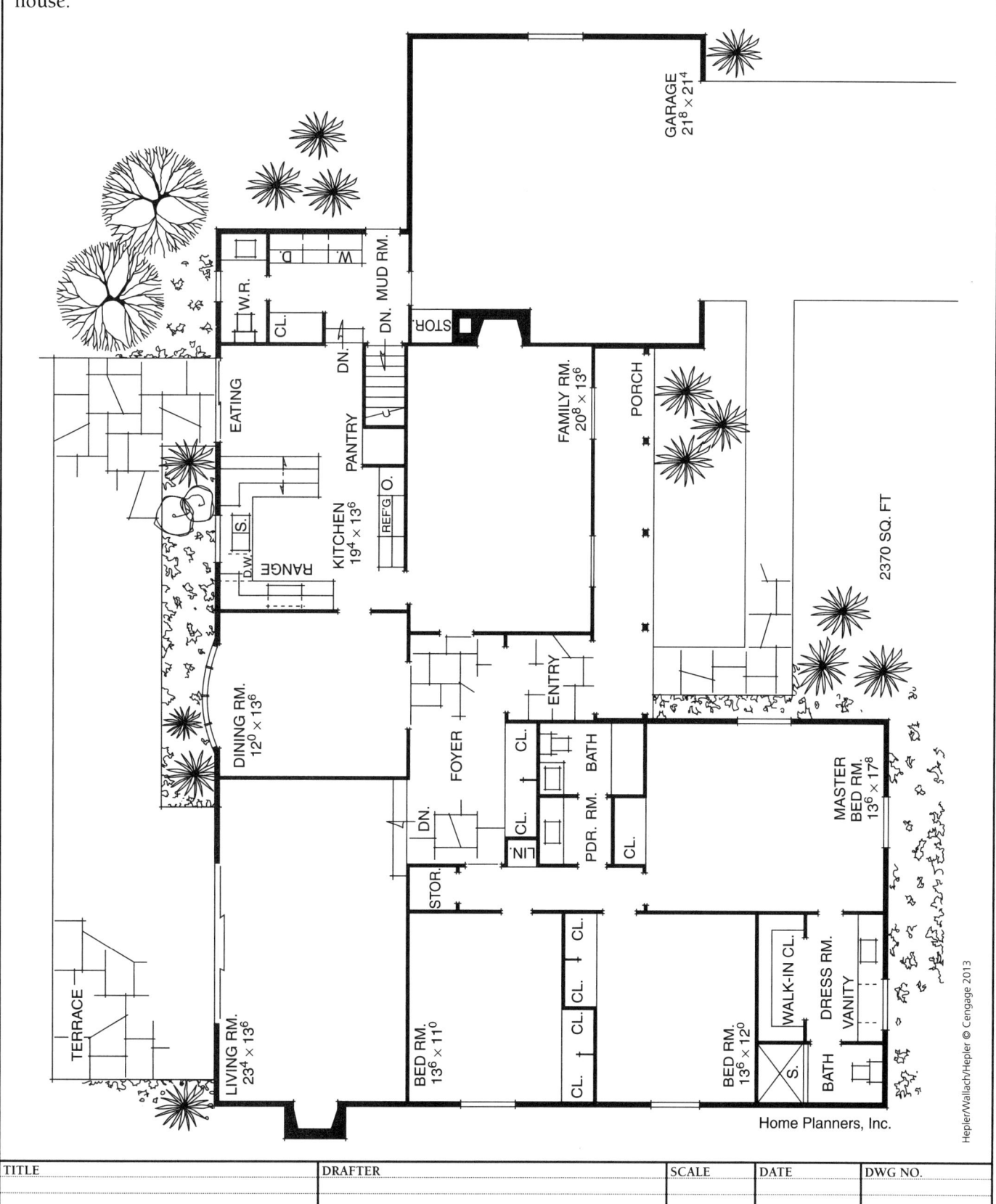

TITLE	DRAFTER	SCALE	DATE	DWG NO.

EXERCISE 31-4

Using the appropriate symbols and line conventions, complete the electrical plan for this unusual home.

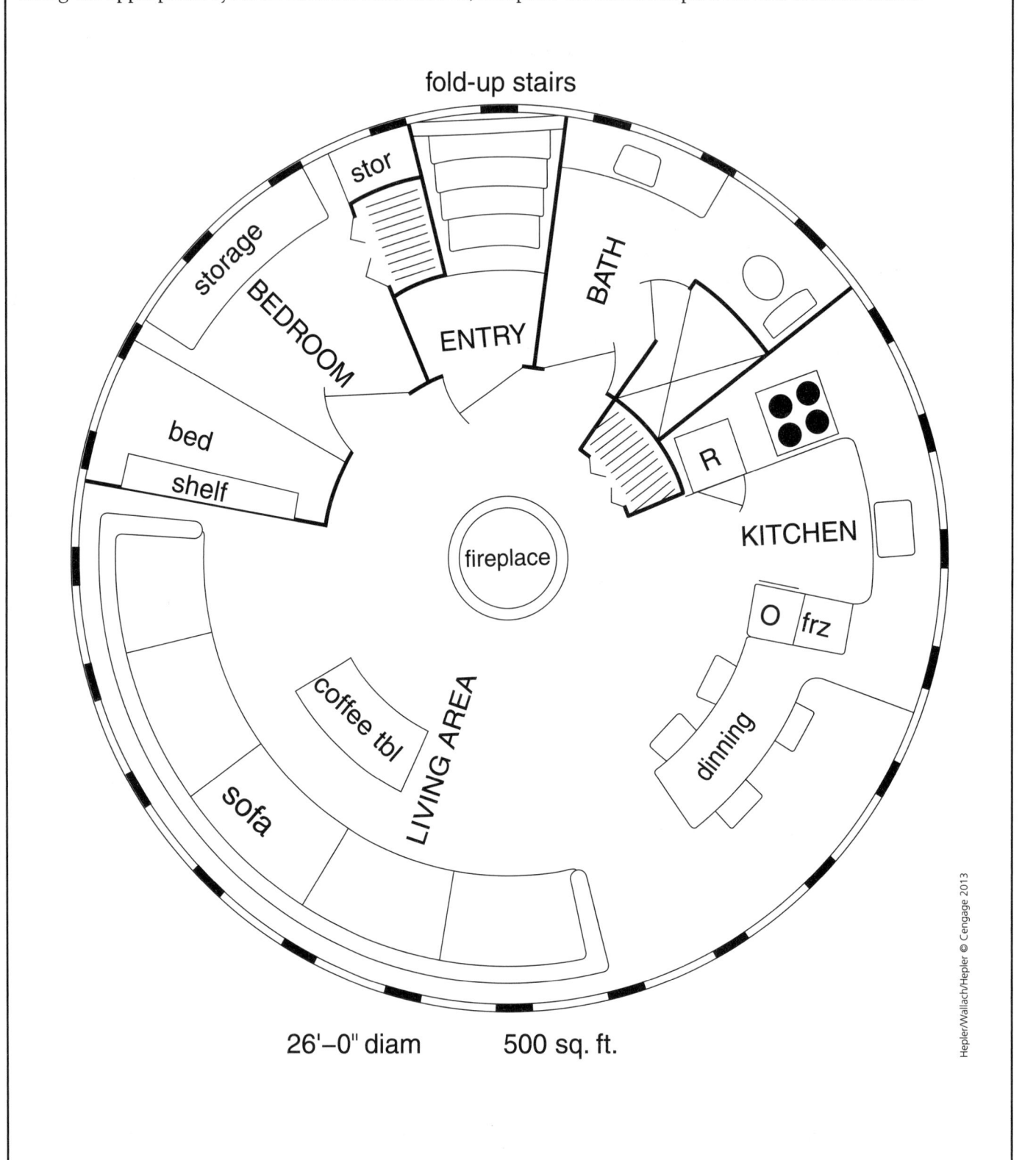

fold-up stairs

stor
storage
BEDROOM
ENTRY
BATH
bed
shelf
R
KITCHEN
fireplace
O frz
coffee tbl
LIVING AREA
dinning
sofa

26'-0" diam   500 sq. ft.

TITLE	DRAFTER	SCALE	DATE	DWG NO.

EXERCISE 31-5

Draw and label each electrical fixture below.

ELECTRICAL SYMBOLS/MISC

Symbol	Label	Symbol	Label
◯	CEILING FIXTURE	⊢◯ (240V)	240 V OUTLET
◯ PS	PULL CORD SWITCH	⊢◯	120 V DOUBLE OUTLET
⊢◯	WALL FIXTURE	⊢◯	120 V TRIPLE OUTLET
⊢©	CLOCK	⊢● DRY.	SPECIAL PURPOSE OUTLET (240 V DRYER)
▨	POWER PANEL	⊢◯ WP	WEATHERPROOF OUTLET
◁	TELEPHONE OUTLET	⊢◯ GFCI	GROUND-FAULT CIRCUIT INTERRUPTER
◀	TELEPHONE JACK	⊙	FLOOR OUTLET
☐▷	BELL	S	SWITCH (TWO WIRES)
Ⓜ	METER	S_3	SWITCH (THREE WIRES)
⊢G	GAS OUTLET	S_4	SWITCH (FOUR WIRES)
⊠	HEAT OUTLET	S_D	AUTOMATIC DOOR SWITCH
☐ FD	FLOOR DRAIN	S_P	SWITCH WITH PILOT LIGHT
Ⓦ H	WATER HEATER	S̲	LOW-VOLTAGE SWITCH

TITLE	DRAFTER	SCALE	DATE	DWG NO.

EXERCISE 32-1

Using the appropriate symbols and line conventions, add a forced-air heating system to this floor plan.

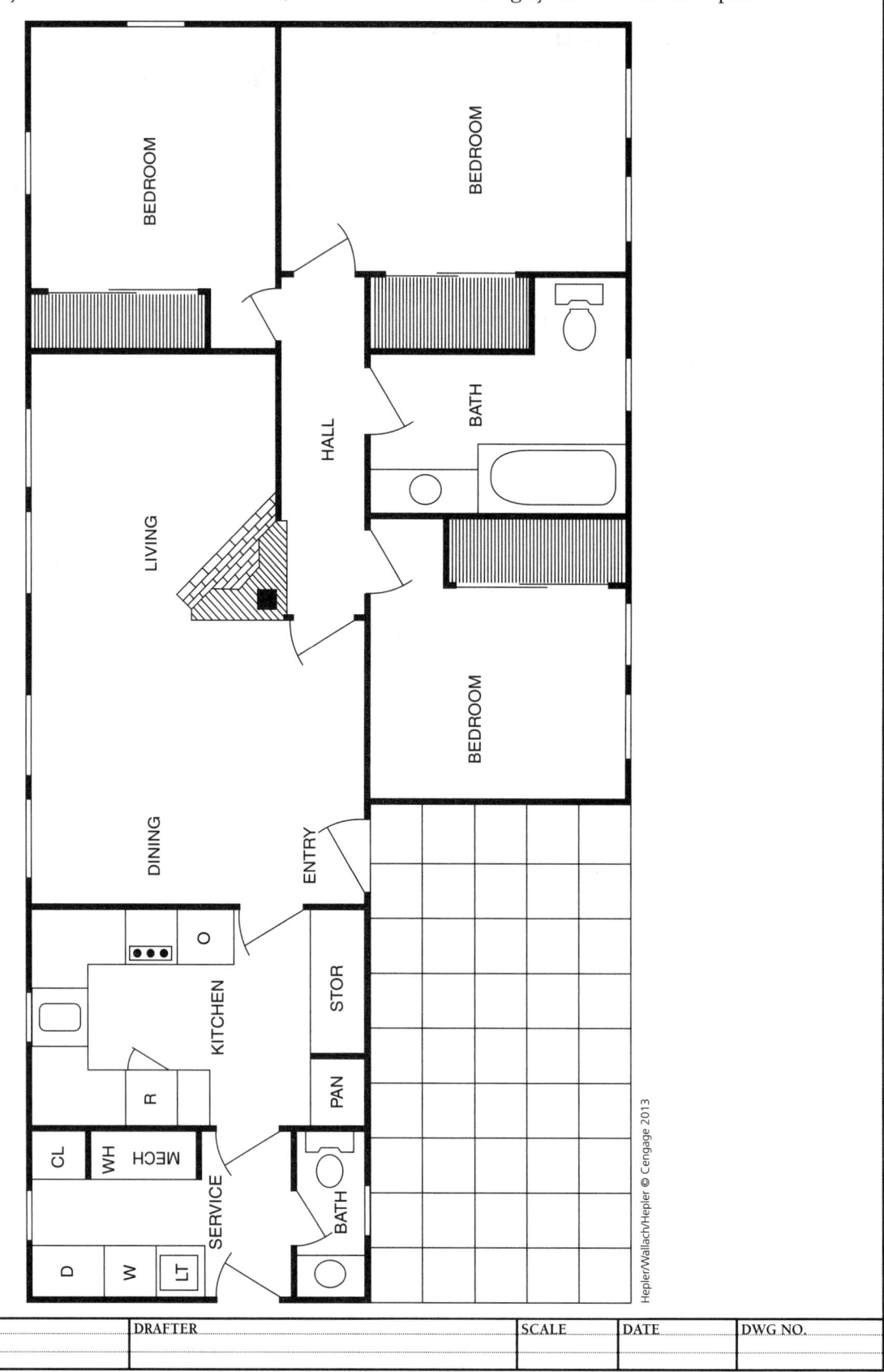

TITLE	DRAFTER	SCALE	DATE	DWG NO.

EXERCISE 32-2

Using the appropriate symbols and line conventions, design a heating system for this house, showing positions of outlets, furnaces, and ducts.

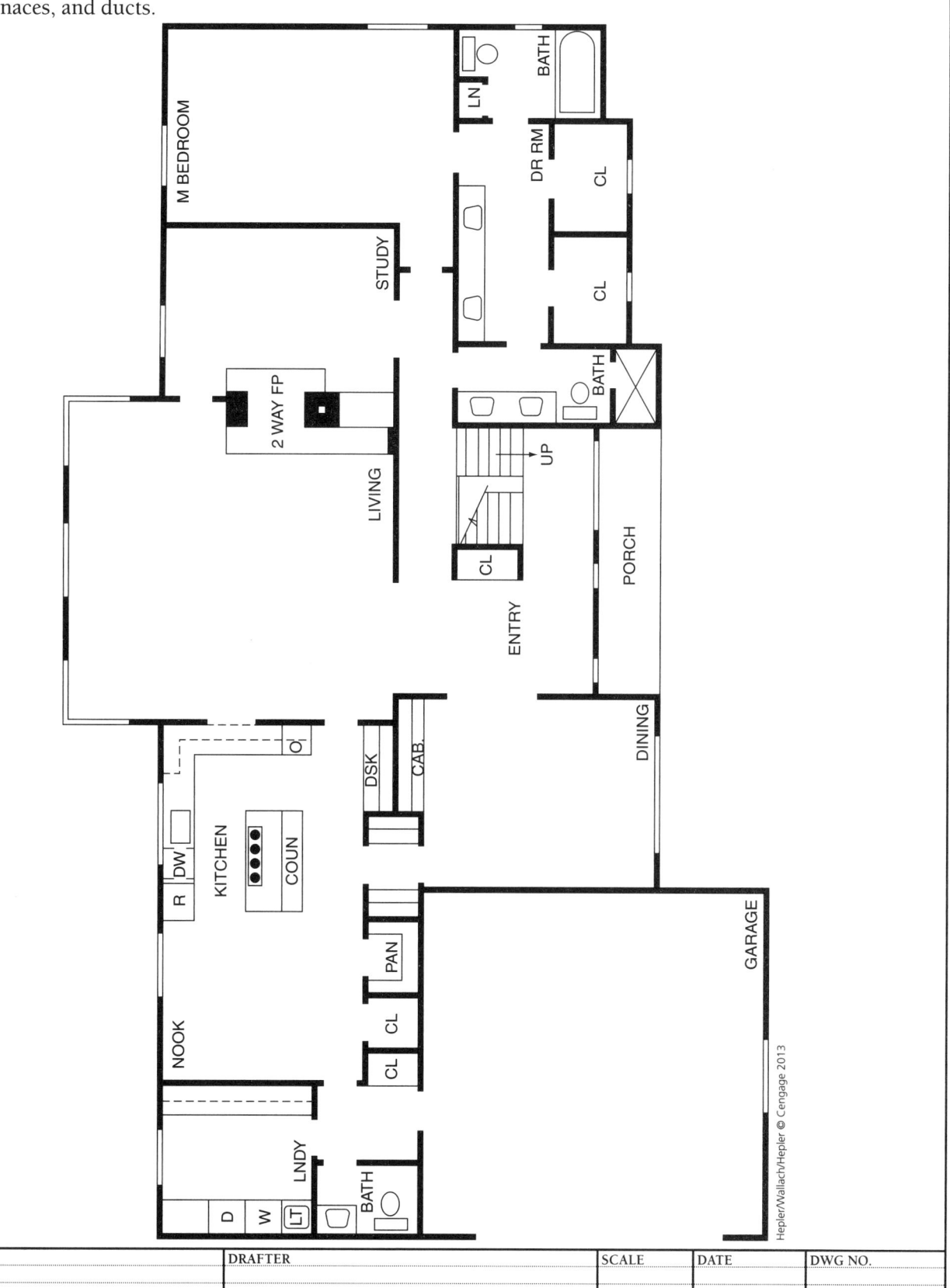

EXERCISE 32-3, PART A

On the elevation, sketch and label a block diagram for an active solar energy system.

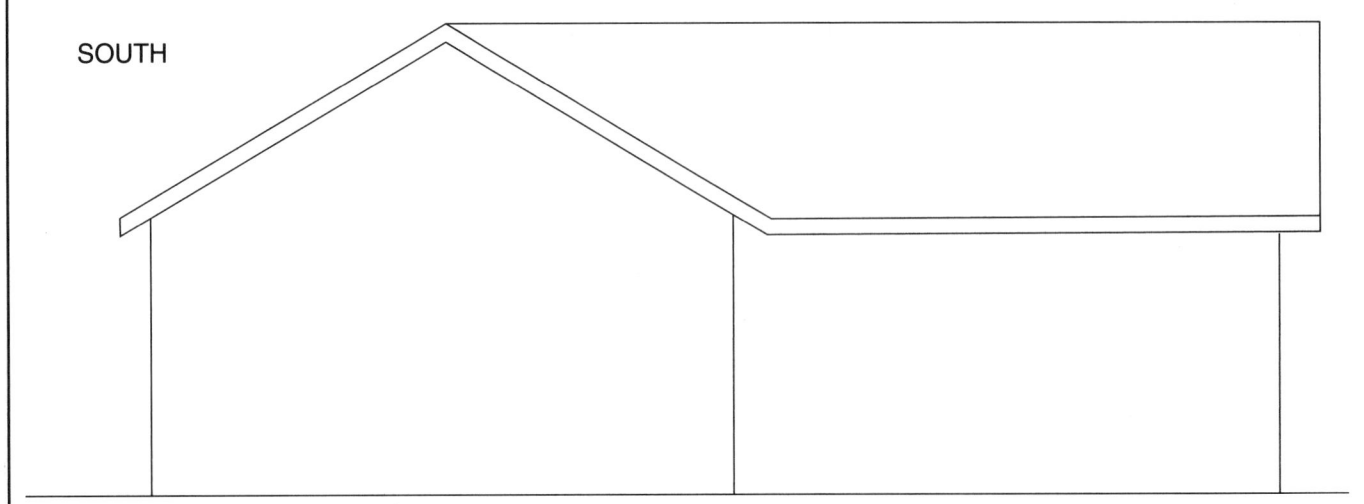

SOUTH

EXERCISE 32-3, PART B

On the floor plan, design and draw a warm-air extended-plenum heating system.

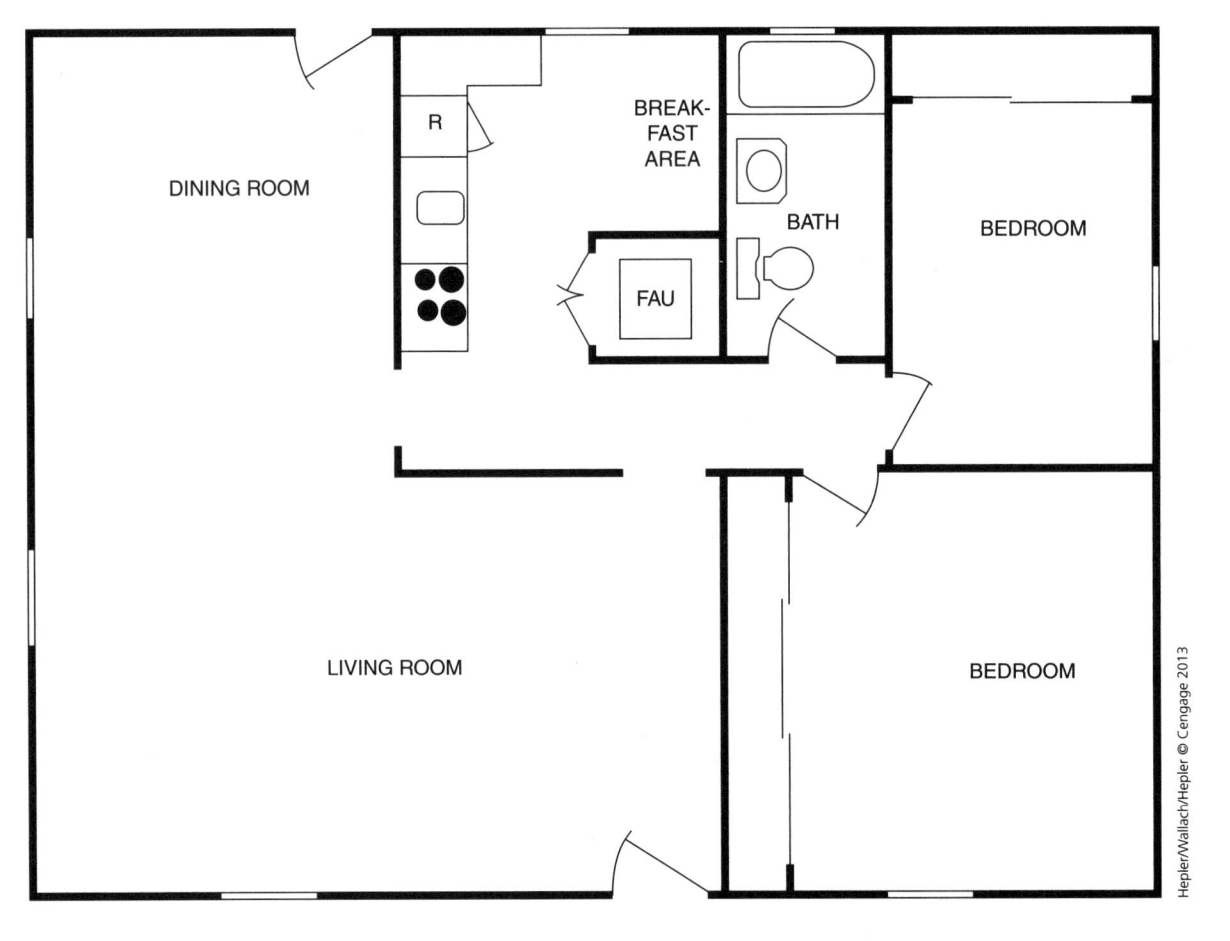

TITLE	DRAFTER	SCALE	DATE	DWG NO.

EXERCISE 32-4

Draw a floor plan for a house of your own design that includes a perimeter forced-air unit (FAU) heating system.

TITLE	DRAFTER	SCALE	DATE	DWG NO.

EXERCISE 33-1

Complete the plumbing plan for the floor plan and elevation.

2" VENT

EXERCISE 33 Plumbing Drawings ■ **159**

EXERCISE 33-2

Complete the plumbing plan for the floor plan and elevation.

TITLE	DRAFTER	SCALE	DATE	DWG NO.

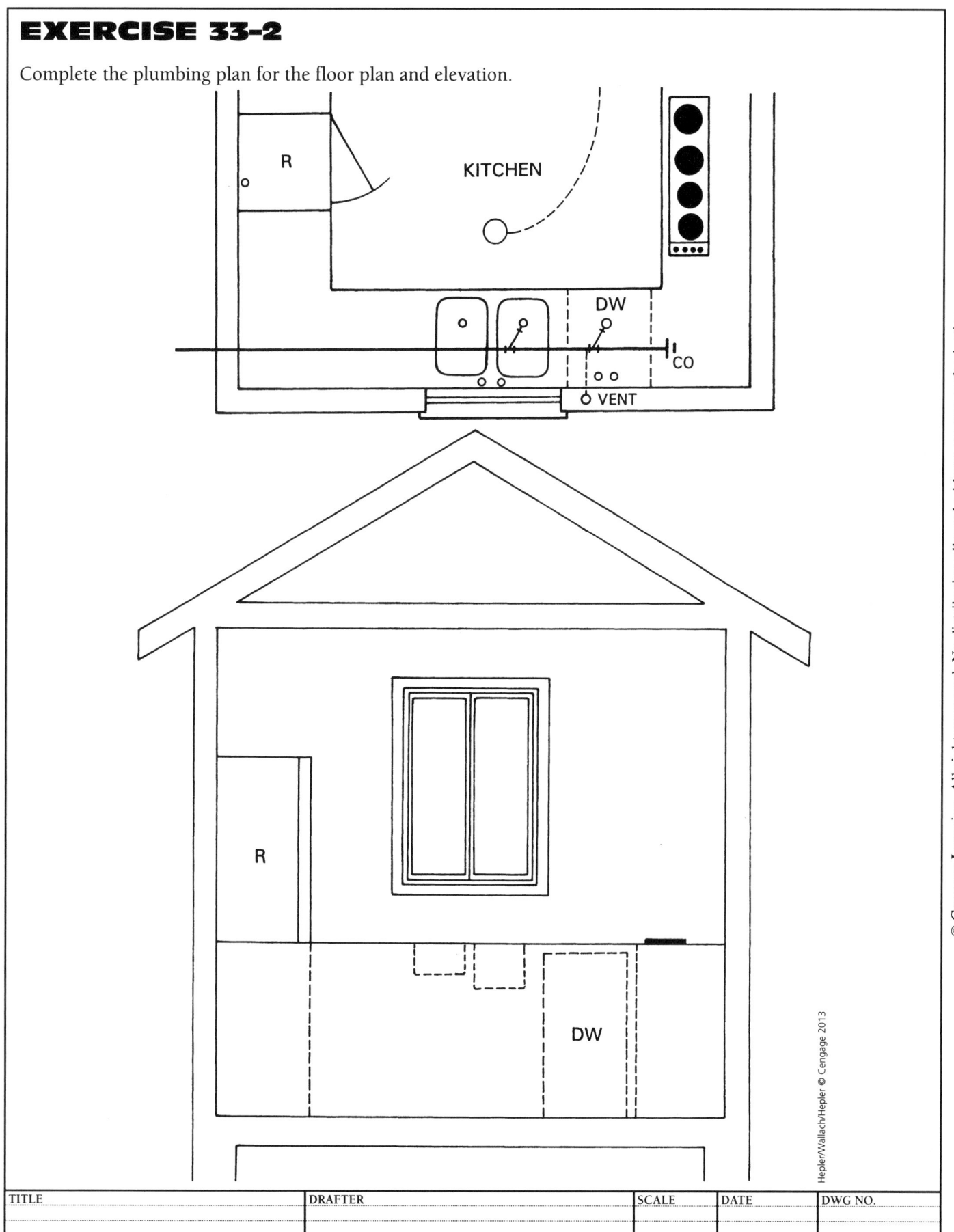

EXERCISE 33-3

Using red and blue pencils, draw a plumbing plan on this drawing.

TITLE	DRAFTER	SCALE	DATE	DWG NO.

EXERCISE 34 Drawing Coordination and Checking ■ **161**

EXERCISE 34-1

The drawing below contains some errors. Find the errors and circle them using a colored pen or pencil. Then note how they should be corrected. Drawing is not to scale (NTS).

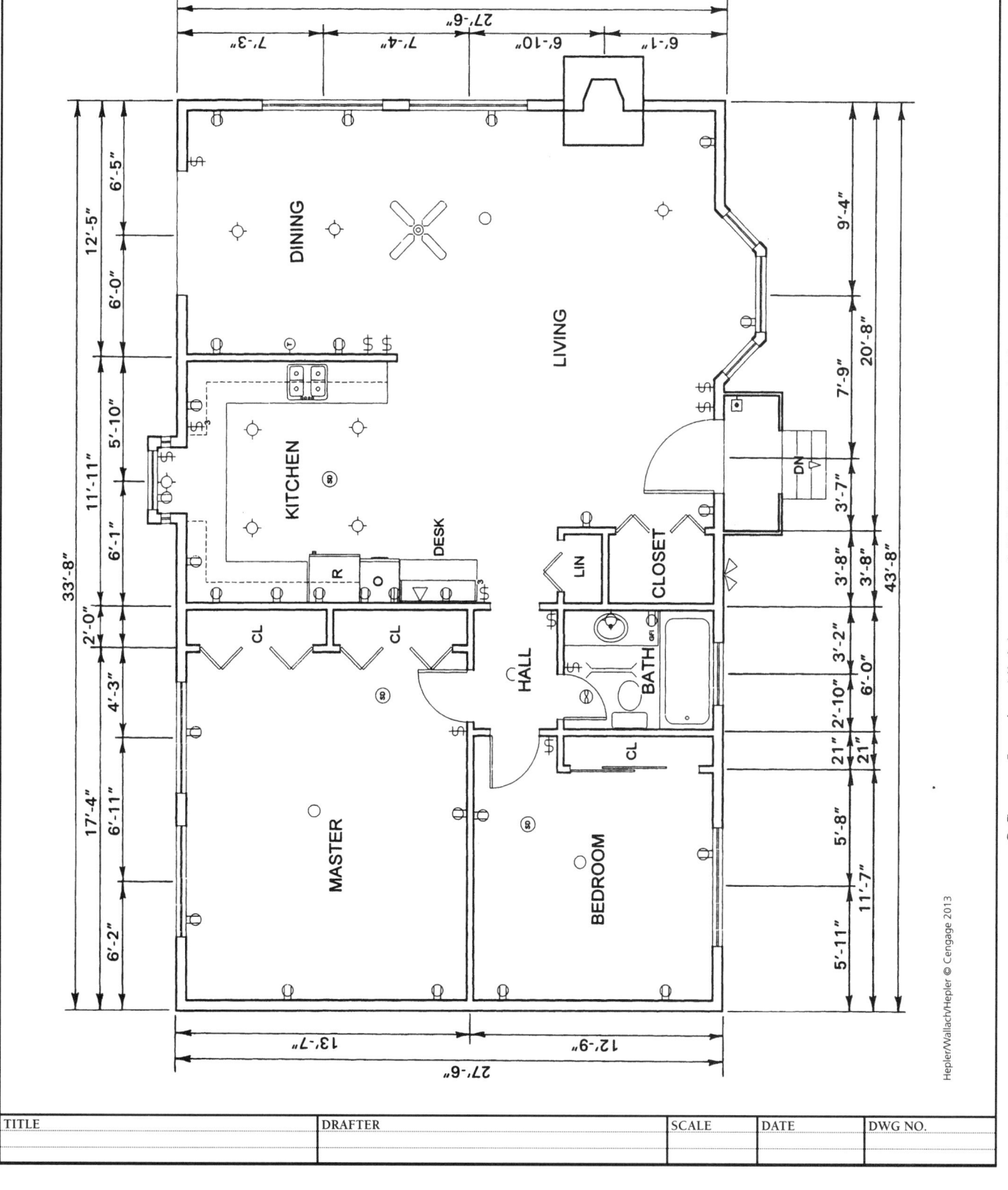

TITLE	DRAFTER	SCALE	DATE	DWG NO.

EXERCISE 34-2

Using the grid, draw a bathroom of your own design, including plumbing and electrical fixtures and connections. When you are finished, exchange drawings with a classmate. Check one another's drawings for mistakes.

| TITLE | DRAFTER | SCALE | DATE | DWG NO. |

EXERCISE 34-3

Geometric symbol problem:

Geometric symbols are found on the following set of plans. Find the location and draw each geometric symbol on the other drawings in the set. Some symbols have been located in several drawings, but not all. Each drawing should contain at least one symbol. Symbols include gray triangles, squares, crosses, circles, rectangles, open circles, 8-point stars, hexagons, and diamonds.

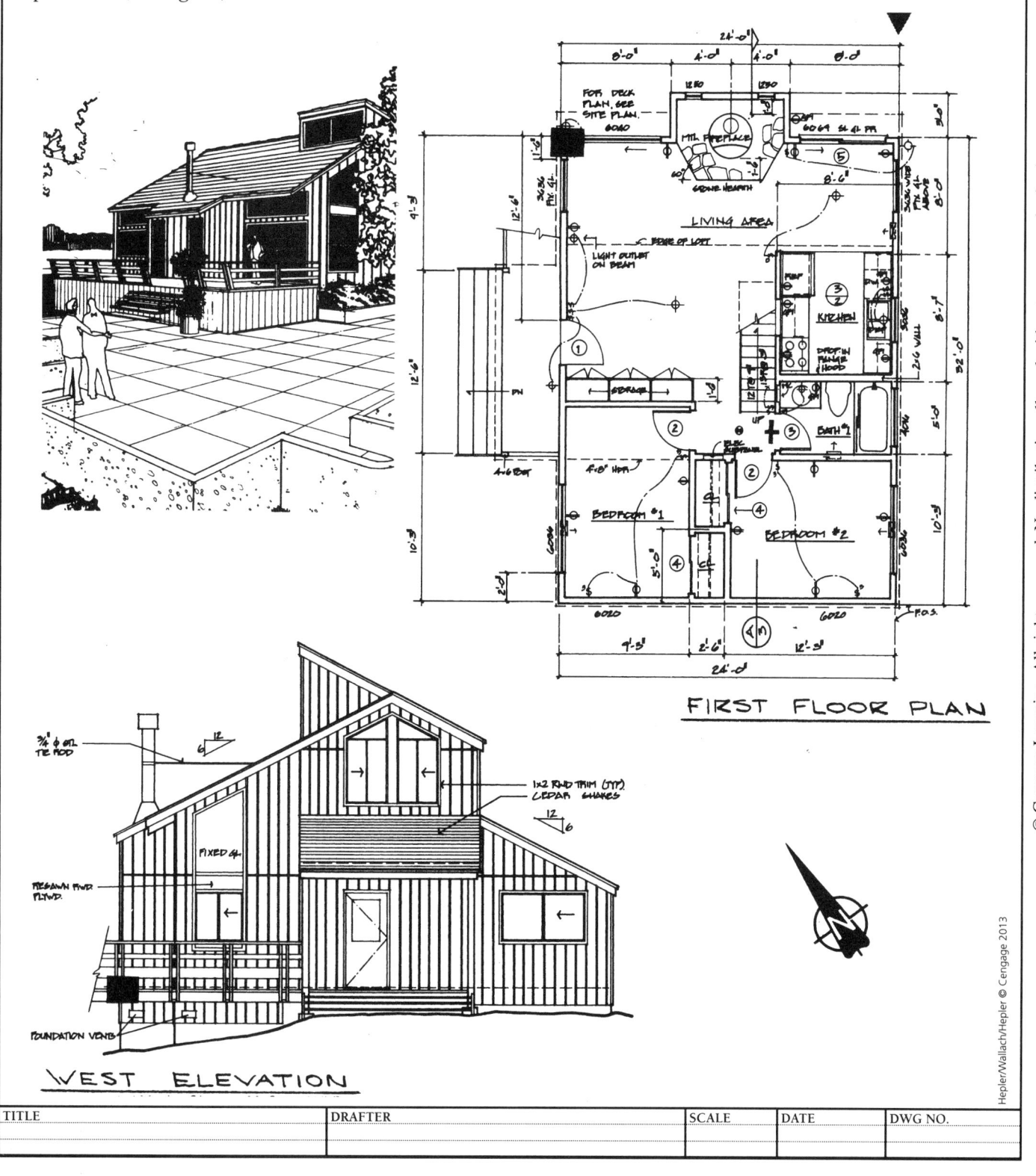

FIRST FLOOR PLAN

WEST ELEVATION

TITLE	DRAFTER	SCALE	DATE	DWG NO.

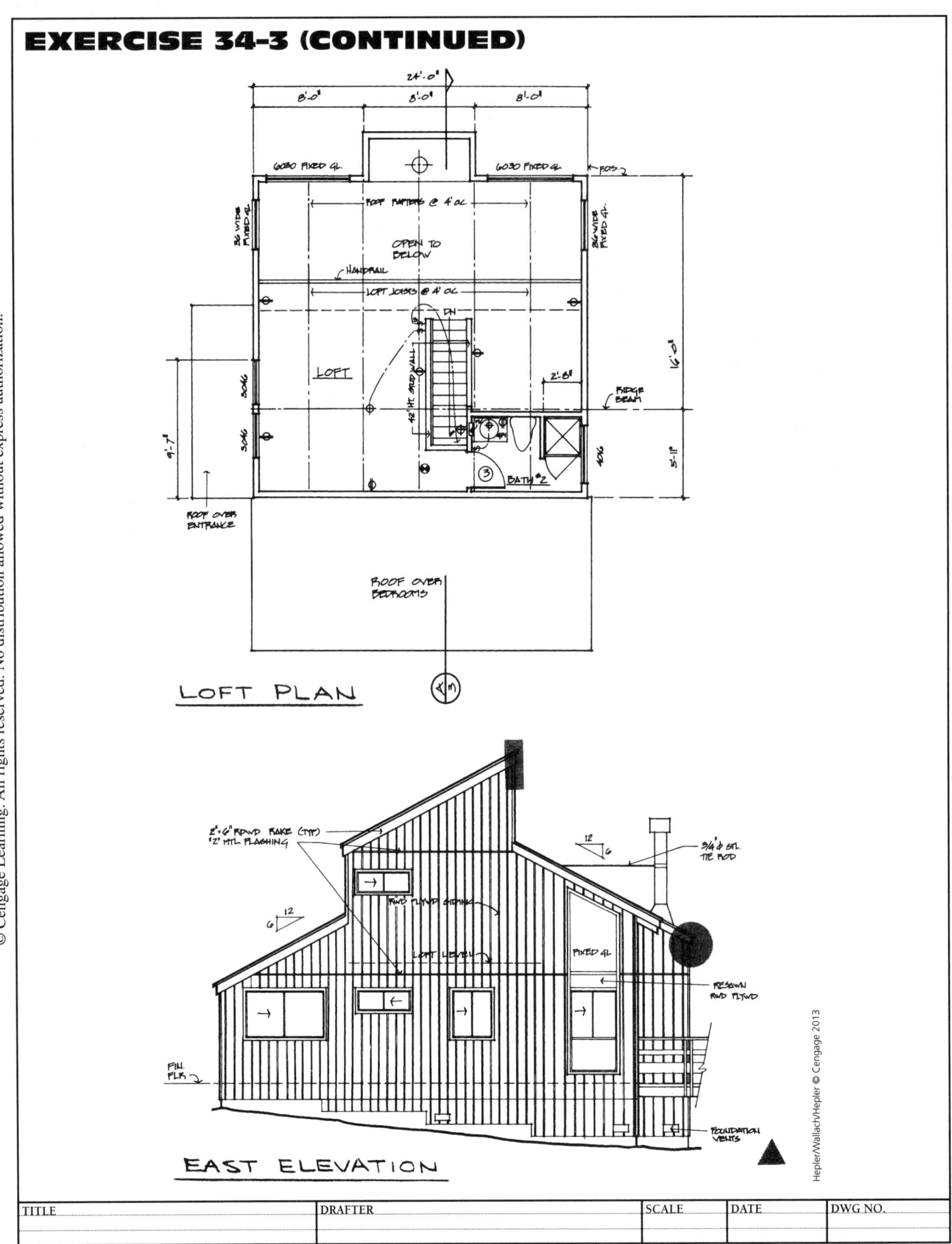

EXERCISE 34-3 (CONTINUED)

LOFT PLAN

EAST ELEVATION

TITLE	DRAFTER	SCALE	DATE	DWG NO.

EXERCISE 34-3 (CONTINUED)

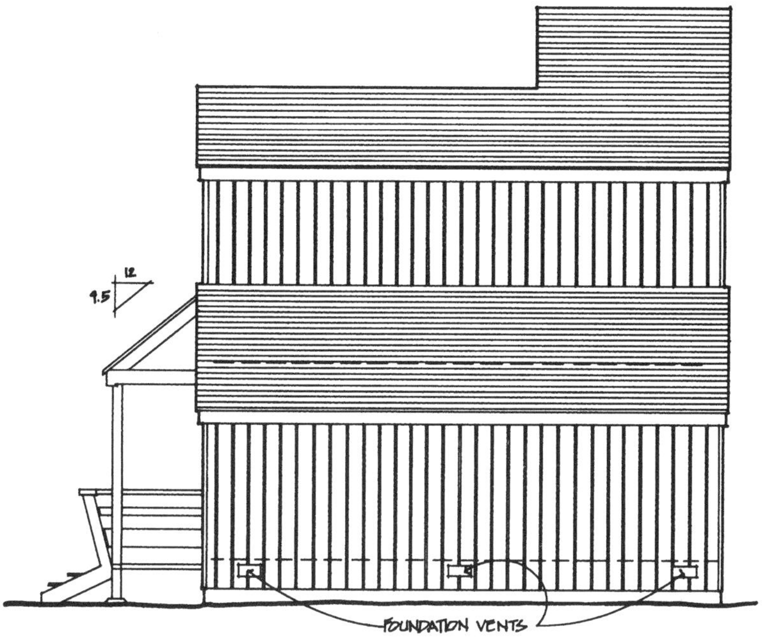

SOUTH ELEVATION

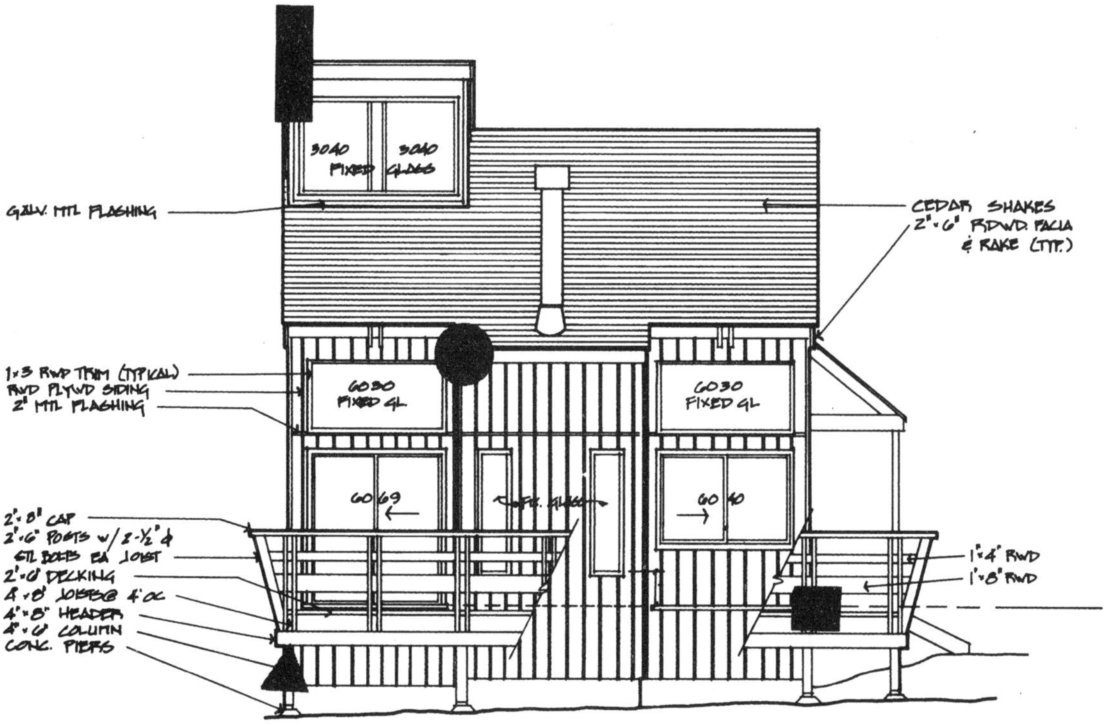

NORTH ELEVATION

Hepler/Wallach/Hepler © Cengage 2013

TITLE	DRAFTER	SCALE	DATE	DWG NO.

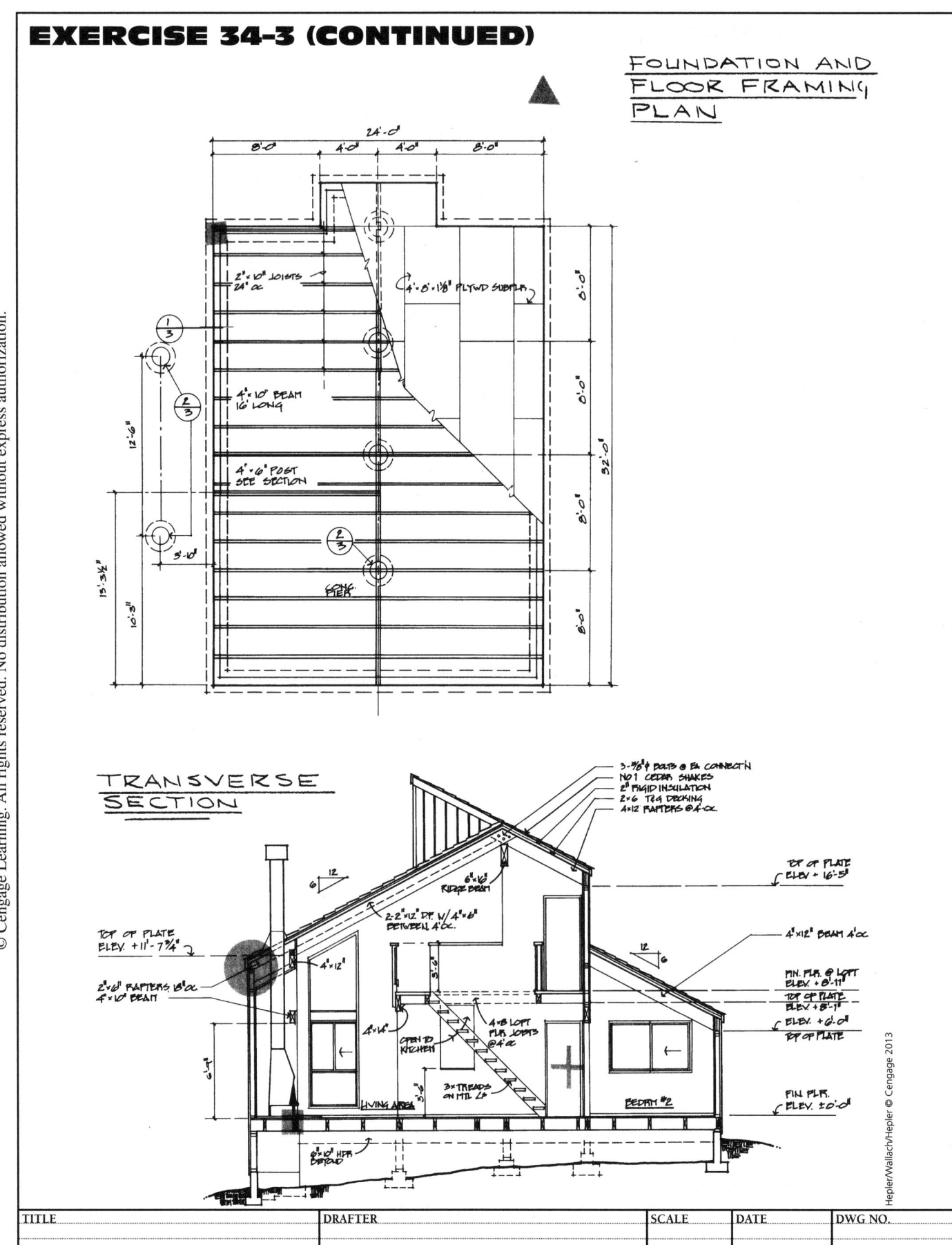

EXERCISE 34-3 (CONTINUED)

FOUNDATION AND FLOOR FRAMING PLAN

TRANSVERSE SECTION

TITLE	DRAFTER	SCALE	DATE	DWG NO.

EXERCISE 34-4

Add a sleeping and living area to the plan in the space below.

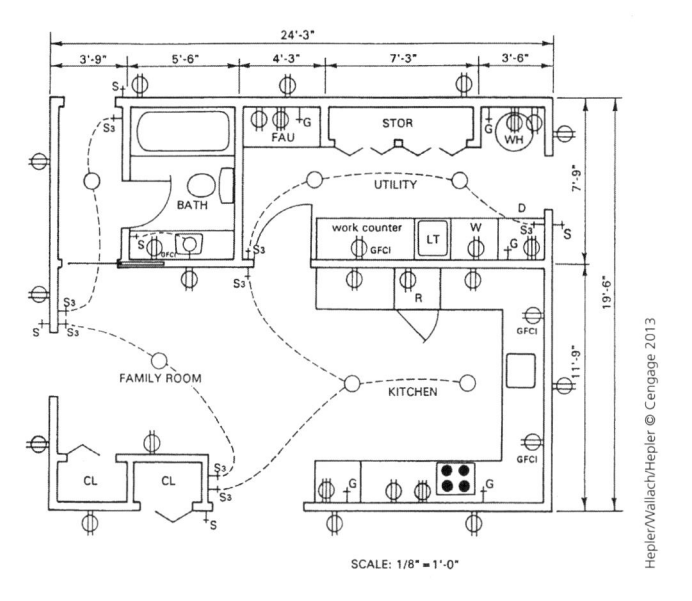

| TITLE | DRAFTER | SCALE | DATE | DWG NO. |

EXERCISE 34-5

Draw an abbreviated floor plan of a house of your design on this property. Draw a profile through the house.

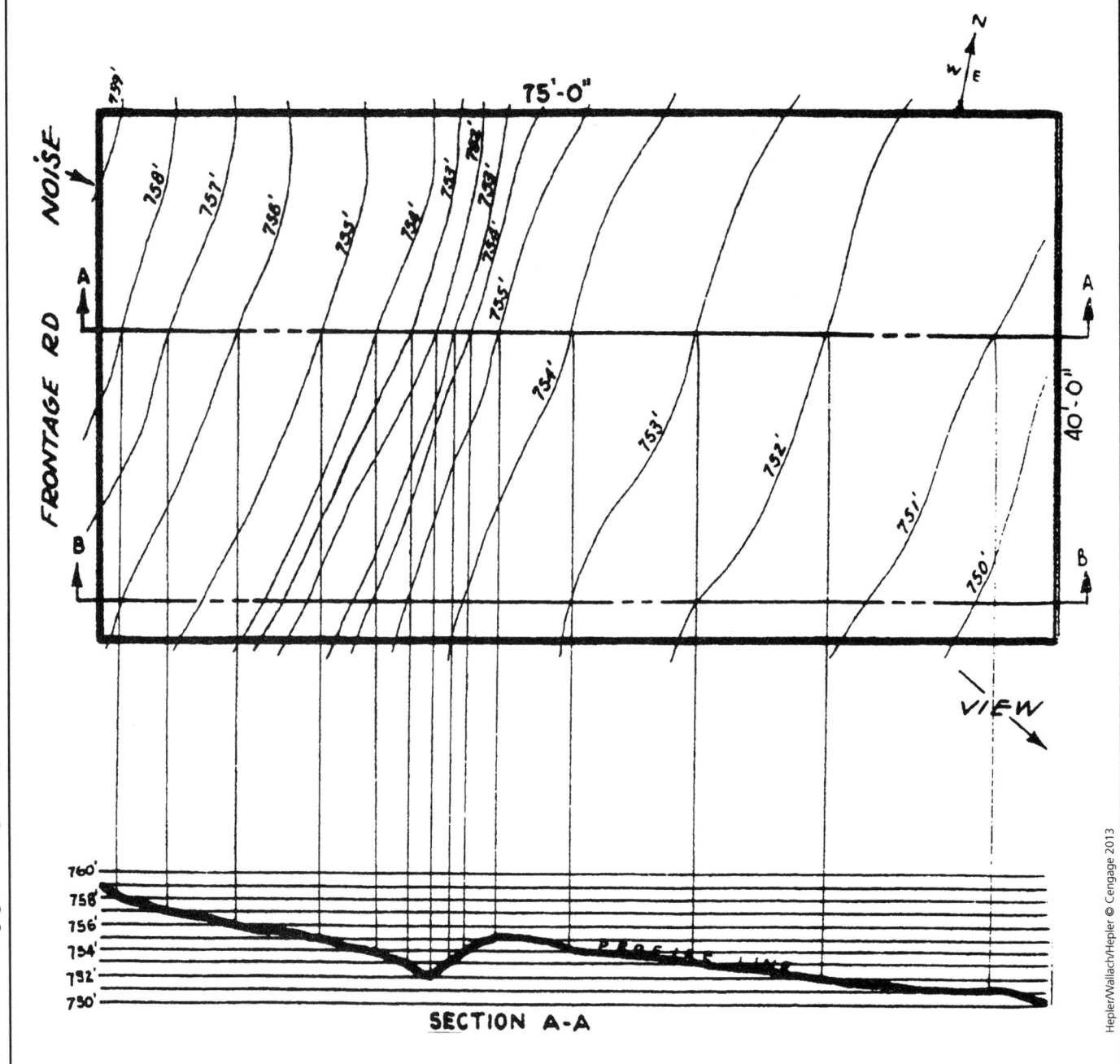

TITLE	DRAFTER	SCALE	DATE	DWG NO.

EXERCISE 35-1

Complete the door and window schedules for the plan shown below.

DOOR SCHEDULE											
Symbol	Width	Height	Thickness	Material	Type	Screen	Quantity	Threshold	Remarks	Manufacturer

WINDOW SCHEDULE									
Symbol	Width	Height	Material	Type	Screen	Quantity	Remarks	Manufacturer	Catalog Number

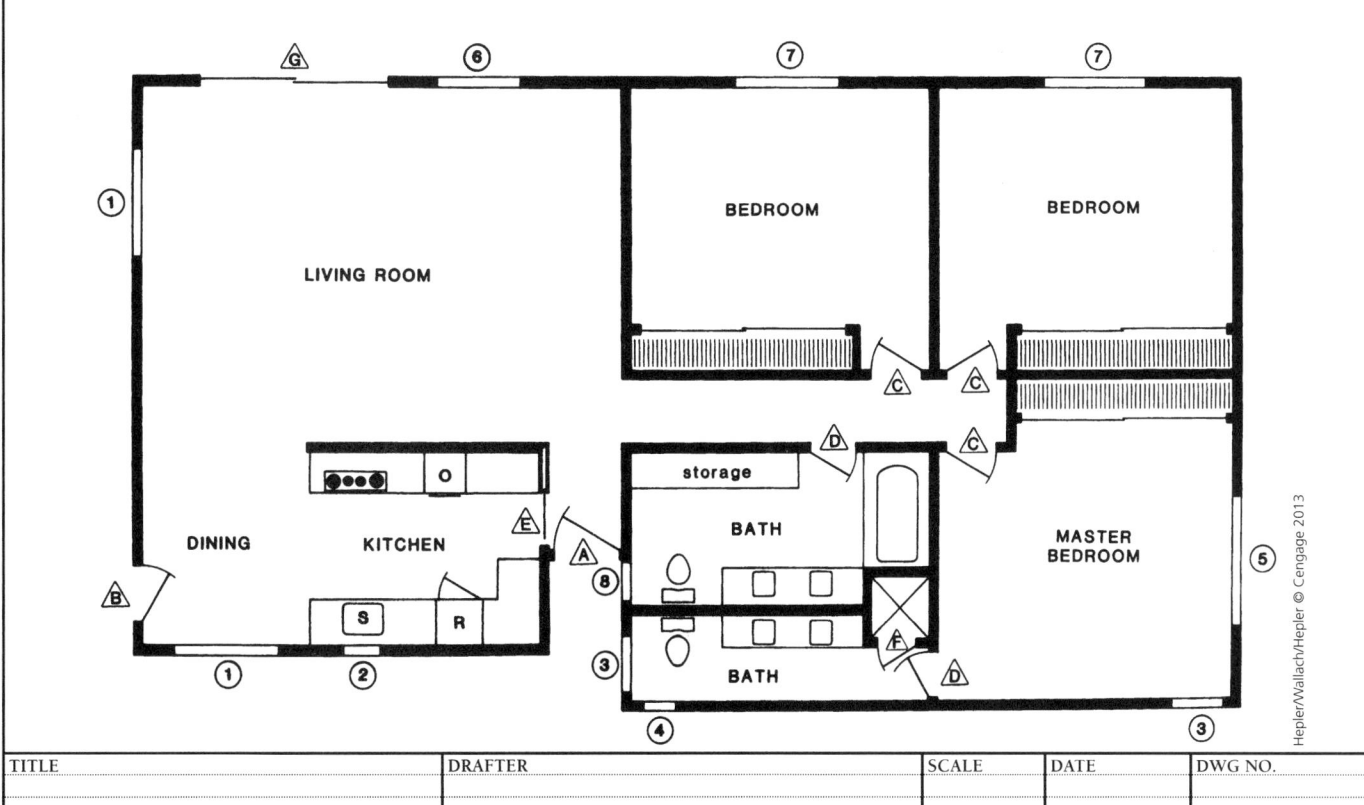

EXERCISE 35-2

Complete the appliance and fixtures schedules for the plan shown below.

APPLIANCE SCHEDULE						
Room	Appliance	Type	Size	Color	Manufacturer	Model Number

FIXTURE SCHEDULE					
Room	Fixture	Type	Material	Manufacturer	Model Number

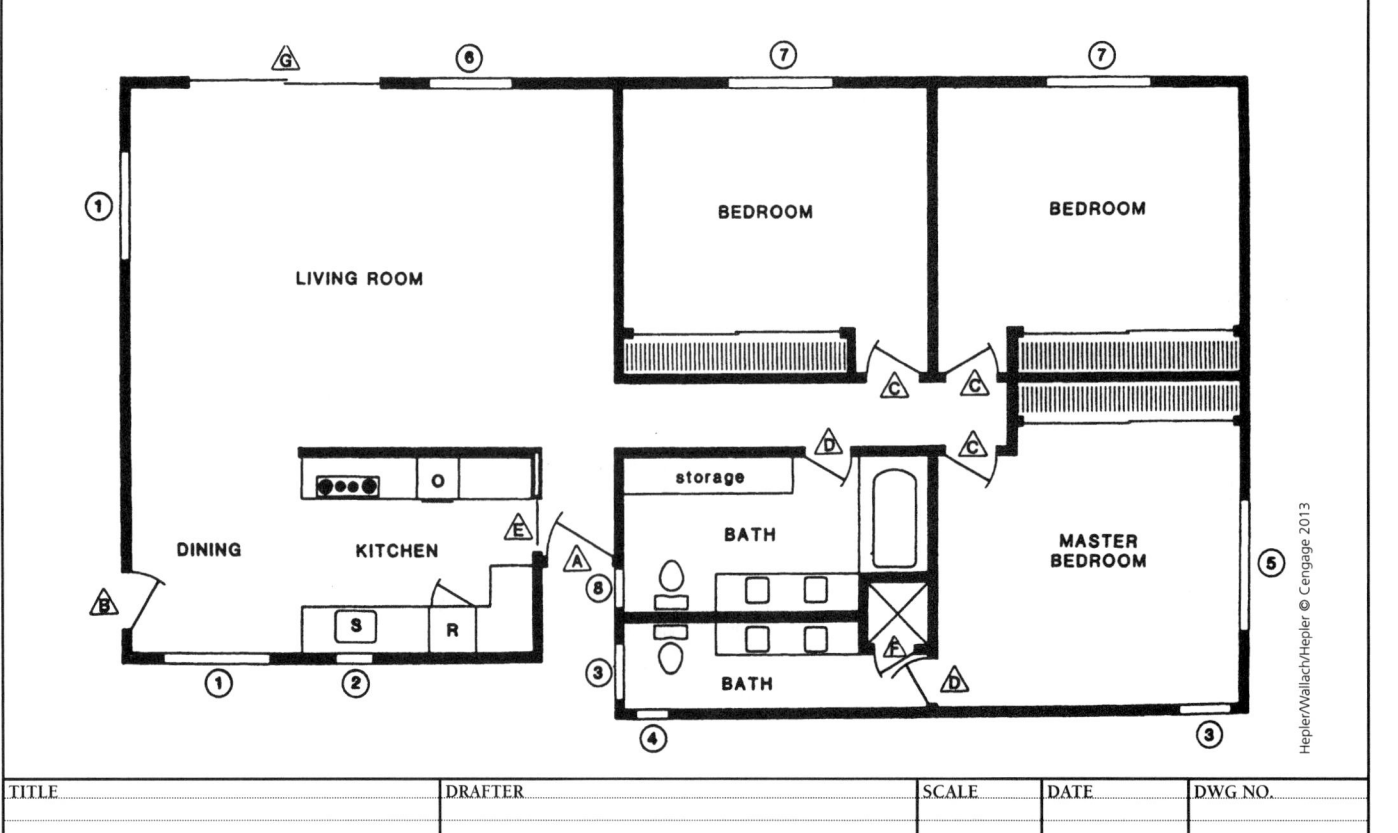

EXERCISE 35-3

Complete the interior design room specification sheet for a room in any plan in the text.

ROOM SPECIFICATION SHEET

SPECIFIC ROOM _____ DATE _____

CLIENT _____ AVAILABLE FUNDS _____

ADDRESS _____

1. ROOM SIZE _____
2. ROOM HEIGHT _____
3. SQUARE FOOTAGE _____
4. FLOOR TREATMENT _____
5. WALL TREATMENT _____
6. CEILING TREATMENT _____
7. WINDOW STYLES _____
8. DOOR STYLES _____
9. PAINTS (colors/location) _____
10. LIGHTING _____
    CEILING _____
    WALLS _____
    LAMPS _____
11. FIXTURES _____
12. APPLIANCES _____
13. ELECTRICAL _____
    SWITCHES _____
    LIGHTING LOCATIONS _____
    SWITCHES _____
    SPECIAL ELECTRICAL EQUIPMENT _____
14. ACCESSORIES _____
15. COLOR SCHEMES _____

NOTES:

TITLE	DRAFTER	SCALE	DATE	DWG NO.

EXERCISE 36-1

Work the problems for each structure.

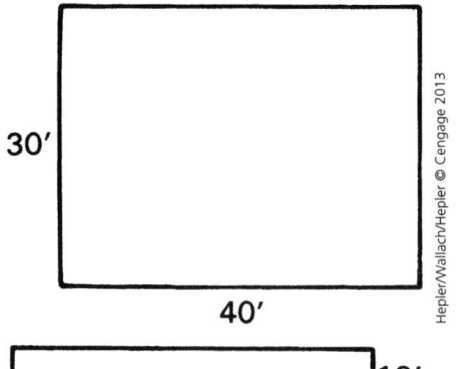

Building costs are $100.00 per square foot.
What is the square footage? _____
What is the total cost? _____

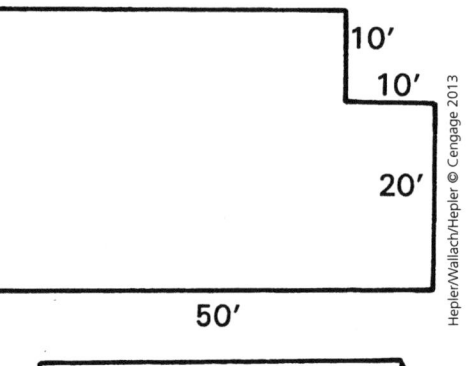

Building costs are $120.00 per square foot.
What is the square footage? _____
What is the total cost? _____

Building costs are $130.00 per square foot.
What is the square footage? _____
What is the total cost? _____

Building costs are $150.00 per square foot.
What is the square footage? _____
What is the total cost? _____

TITLE	DRAFTER	SCALE	DATE	DWG NO.

EXERCISE 36-2

Work the following problems. Write your answers in the spaces provided.

1. About 40% of the cost of a building is for materials. If a house costs $150,000 to build, how much was paid for materials?

    $_____ for materials.

2. How much was left for labor, real estate, and other costs?

    $_____ for labor, real estate, and other costs.

3. About 40% of the cost of a building is for labor. The Sanchez family spent $160,000 for labor. How much did their house cost altogether?

    $_____ Total cost.

4. Calculate the cubic feet in the following structures:

    Height 12′; length 45′; width 27′–6″ = _____ cu ft

    Length 67′–6″; height 9′, width 35′ = _____ cu ft

5. The Wilsons want to buy a house that costs $200,000. They want to make a down payment of 15%. How much will they need?

    $_____ down payment.

6. The Chin family bought a home for $237,240 with a 15-year mortgage. Excluding interest, how much was each monthly payment?

    $_____ per month.

7. The Chaudhuri family is considering a 30-year mortgage. At 9% interest, their payments would be $876. At 13.5%, their payments would be $1,285. What percent of increase does the higher payment represent?

    _____ percent increase.

8. Beth pays $1,350 each year in property taxes. Taxes are figured at $1.35 per $100 of the property's value. What is the value of her property?

    $_____ total value.

TITLE	DRAFTER	SCALE	DATE	DWG NO.

EXERCISE 37-1

Using the table of building-code ceiling joist spans, add the correct notations and directions to the drawing. (*Note:* Drawing is not to scale.)

BUILDING CODE SPECIFICATIONS FOR CEILING JOIST SPANS		
SIZE	JOIST SPACING	SPAN
2 x 4	12"	10'–0"
	16"	9'–0"
	24"	8'–0"
2 x 6	12"	16'–0"
	16"	14'–0"
	24"	12'–6"
2 x 8	12"	21'–6"
	16"	10'–0"
	24"	10'–0"

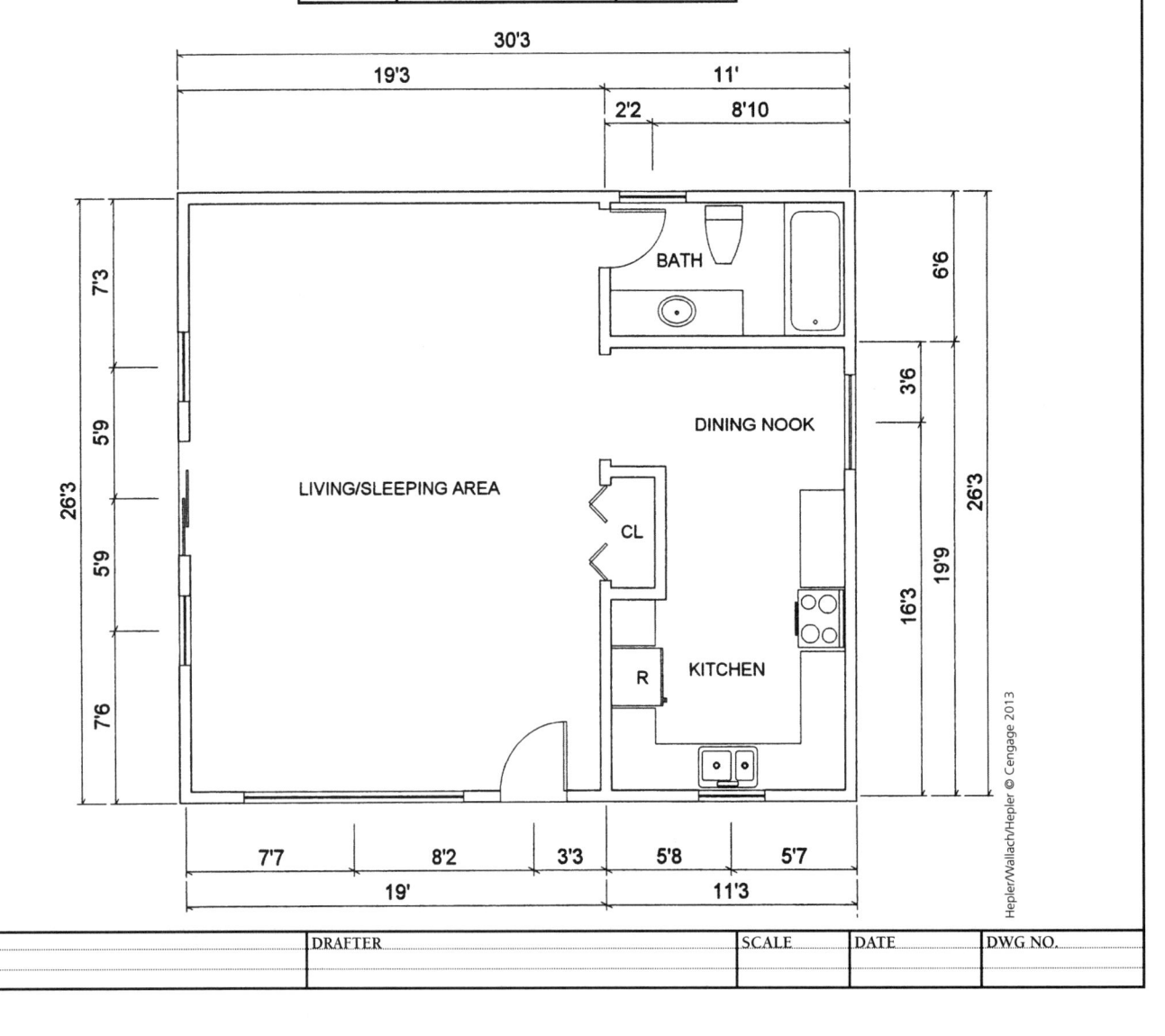

EXERCISE 37-2

Using the table of building-code residential rafter spans, complete this drawing by adding framing for a gable roof having a pitch of 8:12. (*Note:* Drawing is not to scale.)

\multicolumn{8}{c}{**BUILDING CODE SPECIFICATIONS FOR RESIDENTIAL RAFTER SPANS**}
SIZE	RAFTER SPACING	LESS THAN 4:12 PITCH	MORE THAN 4:12 PITCH	SIZE	RAFTER SPACING	LESS THAN 4:12 PITCH	MORE THAN 4:12 PITCH
2 x 4	12"	9'–0"	10'–0"	2 x 10	12"	23'–0"	25'–0"
	16"	8'–0"	8'–6"		16"	21'–0"	22'–6"
	24"	6'–6"	7'–0"		24"	17'–6"	19'–0"
	32"	5'–6"	6'–0"		32"	15'–0"	16'–6"
2 x 6	12"	14'–0"	16'–0"	2 x 12	12"	27'–0"	29'–0"
	16"	12'–6"	13'–6"		16"	25'–0"	27'–0"
	24"	10'–6"	11'–0"		24"	21'–0"	23'–6"
	32"	9'–0"	9'–0"		32"	19'–0"	21'–6"
2 x 8	12"	19'–0"	21'–6"				
	16"	17'–0"	18'–6"				
	24"	13'–6"	15'–0"				
	32"	11'–6"	13'–0"				

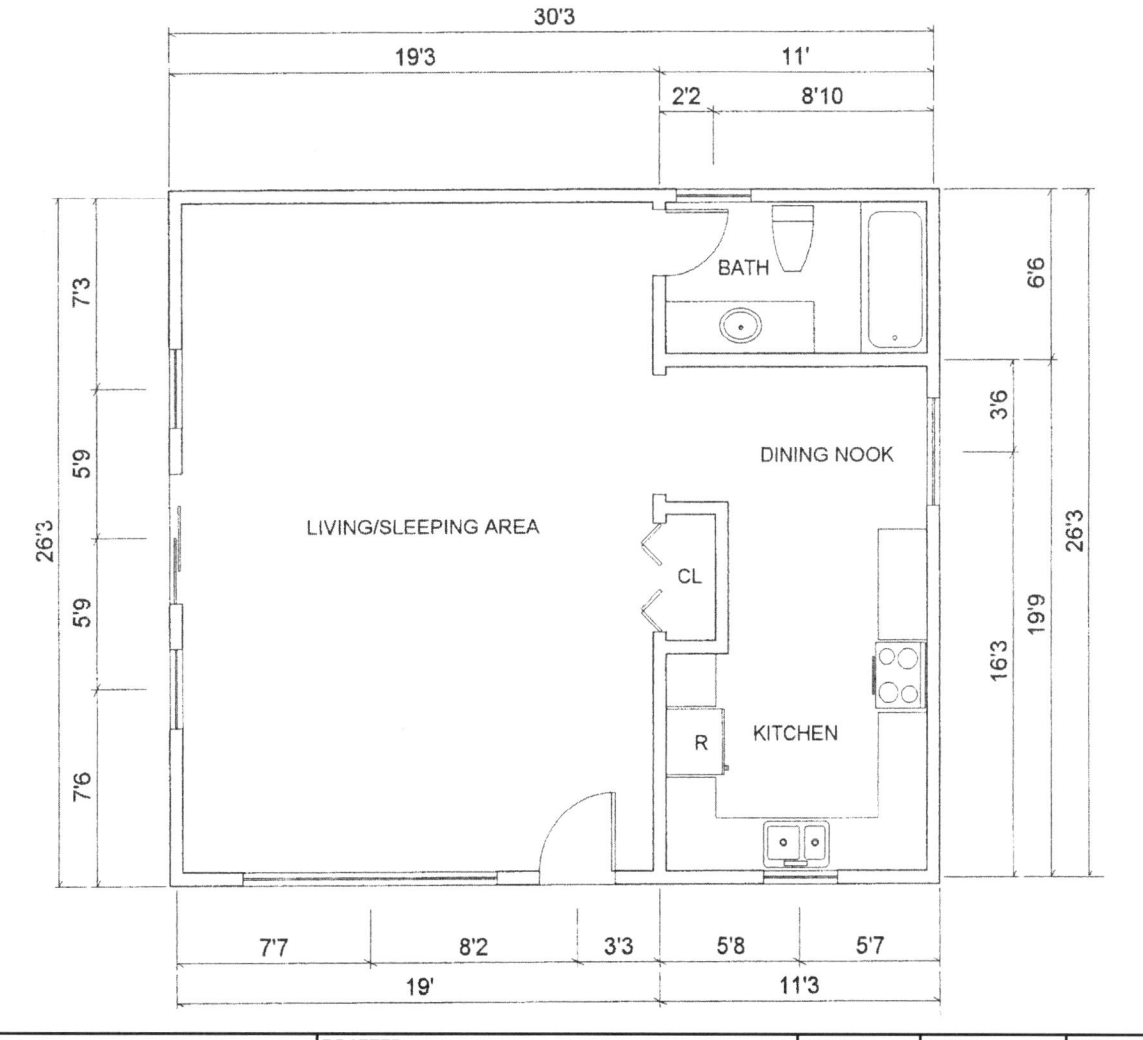

EXERCISE 37-3

Using the table of building-code floor joist spans, complete the plan for a T-foundation floor system, including a girder and piers. (Use a scale of 1/8" = 1'–0".)

\multicolumn{6}{c}{BUILDING CODE SPECIFICATIONS FOR FLOOR JOIST SPANS}
SIZE	JOIST SPACING	SPAN	SIZE	JOIST SPACING	SPAN
2 x 4	12"	10'–0"	2 x 10	12"	16'–0"
	16"	9'–0"		16"	15'–0"
	24"	7'–0"		24"	12'–0"
2 x 6	12"	13'–0"	2 x 12	12"	20'–0"
	16"	12'–0"		16"	18'–0"
	24"	10'–6"		24"	15'–0"
2 x 8	12"	14'–0"	2 x 14	12"	23'–0"
	16"	13'–0"		16"	21'–0"
	24"	11'–0"		24"	17'–0"

TITLE	DRAFTER	SCALE	DATE	DWG NO.

EXERCISE 37-4

Using the table, draw section details for a T foundation and piers, including part of the floor framing.

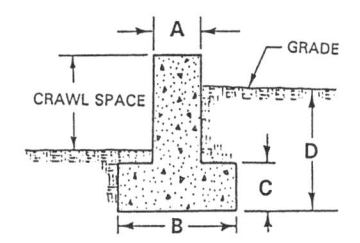

UNIFORM BUILDING CODE FOR T FOUNDATIONS				
BUILDING HEIGHTS	A–WIDTH OF OF WALL	B–WIDTH OF FOOTING	C–HEIGHT OF FOOTING	D–DEPTH OF OF FOOTING
ONE STORY	6"	12"	6"	12"
TWO STORIES	8"	15"	7"	18"
THREE STORIES	10"	18"	8"	24"

Girder spans (pier spacing) for non-bearing walls is 3'–0".
Girder spans for bearing walls is 5'–6".
All calculations are for Douglas fir #2 grade (D.F. #2).

TITLE	DRAFTER	SCALE	DATE	DWG NO.

SECTION 2
Drawing Templates

The following pages contain room and furniture templates that can be used on a floor plan or elevation drawing to plan and test space relationships and distances. The templates are prepared at a scale of 1/4″ = 1′–0″ to be compatible with most residential plans. If different proportional scales are desired the template page can be expanded or reduced by the percentage desired: 1/8″ = 1′–0″, 1/2″ = 1′–0″, etc.

If three-dimensional templates are needed, refer to the exercise in Chapter 21 in this workbook.

T-1: FLOOR PLAN ROOMS

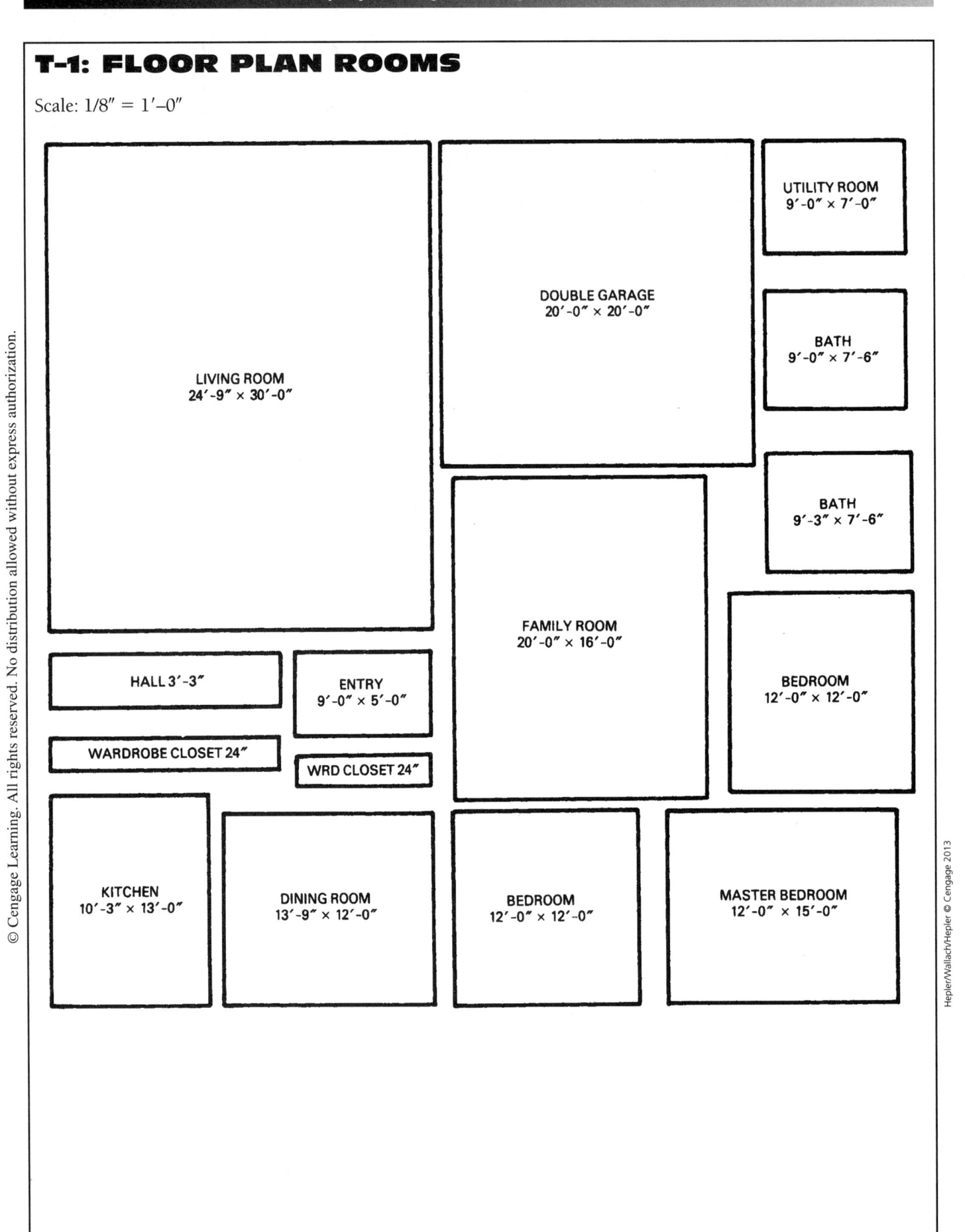

Scale: 1/8" = 1'-0"

LIVING ROOM
24'-9" × 30'-0"

DOUBLE GARAGE
20'-0" × 20'-0"

UTILITY ROOM
9'-0" × 7'-0"

BATH
9'-0" × 7'-6"

BATH
9'-3" × 7'-6"

FAMILY ROOM
20'-0" × 16'-0"

BEDROOM
12'-0" × 12'-0"

HALL 3'-3"

ENTRY
9'-0" × 5'-0"

WARDROBE CLOSET 24"

WRD CLOSET 24"

KITCHEN
10'-3" × 13'-0"

DINING ROOM
13'-9" × 12'-0"

BEDROOM
12'-0" × 12'-0"

MASTER BEDROOM
12'-0" × 15'-0"

T-2: LIVING AREA

Scale: 1/4" = 1'-0"

CLUB CHAIR 30" × 33"

WING CHAIR 36" × 33"

ARM CHAIR 27" × 27"

LARGE SOFA 7'-3" × 3'-3"

SMALL SOFA 4'-6" × 2'-6"

END TABLE 16" × 30"

MODULAR SEATING UNIT 30" × 30"

FOOTSTOOL 27" × 22"

TV CONSOLE 45" × 20"

PORTABLE TV 30" × 18"

BOOKSHELF VARIES, 10"

CHINA CABINET 36" × 15"

BABY GRAND PIANO

PIANO BENCH 14" × 36"

PIANO 27" × 66"

STOOL 10" DIA

STOOL 12" SQ

DINING TABLE 3'-0" × 8'-0"

DINING TABLE 4'-0" SQ

DINING TABLE 5'-3" DIA

PLAY PEN 40" SQ

CORNER CABINET 36"

DINING TABLE 3'-0" × 6'-0"

DINING TABLE 4'-0" DIA

DINING TABLE 3'-0" × 5'-0"

BRIDGE TABLE 3'-0" SQ

BEN FRANKLIN STOVE 39" × 20"

DINING TABLE 36" SQ

TABLE 24" SQ

TABLE 30" DIA

FREESTANDING FIREPLACE 36" DIA

COFFEE TABLE 38" × 18"

SMALL DESK 42" × 18"

LARGE DESK 66" × 36"

FIREPLACE 30"

HEARTH 18", VARIES

BREAKFRONT 20" × 54"

T-2: LIVING AREA (CONTINUED)

- THREE SWING SET 13'-0" × 6'-0"
- PICNIC TABLE 5'-0" DIA
- PLANTERS 24" DIA, 12" DIA
- FISH POND 57" × 40"
- DOG HOUSE 28" × 40"
- HAMMOCK 90" × 12"
- SAUNA 6'-0" × 5'-0" (HTR)
- SLIDE 15" × 100"
- HOT TUB 6'-0" DIA
- LADDER 22", 54", 48", 22", 18"
- UMBRELLA 5'-0" DIA
- BENCH VARIES × 18"
- RECLINING CHAIR 70" × 30"
- PICNIC TABLE 48" × 36"
- PORTABLE BARBEQUE 24" × 16"
- MASONRY BARBEQUE VARIES × 24"
- OUTDOOR CHAIR 22" × 22"
- MOTORCYCLE 72" × 32"
- WAGON 39" × 27"
- TRICYCLE 36" × 24"
- LARGE CAR 6'-9" × 19'-6"
- STANDARD CAR 6'-6" × 17'-6"
- COMPACT CAR 6'-0" × 15'-0"
- BICYCLE
- STORAGE CABINET 18" × VARIES
- WORKBENCH 24" × VARIES

T-3: BEDROOM AND SERVICE AREA

Scale: 1/4" = 1'–0"

- CRIB 25" × 51"
- YOUTH BED 33" × 66"
- COT 30" × 74"
- TWIN BED 39" × 74"
- FULL-SIZE BED 54" × 74"
- QUEEN-SIZE BED 60" × 80"
- KING-SIZE BED 72" × 84"
- WARDROBE CLOSET 24" × VARIES
- CHEST 48" × 18"
- TRIPLE DRESSER 84" × 18"
- NIGHT STAND 15" SQ
- DRESSING TABLE 40" × 18"
- DOUBLE DRESSER 58" × 18"
- ROUND BED 96" DIA
- VANITY 39" × 18"
- BENCH 22" × 12"
- CHAIR 20" × 20"
- UPHOLSTERED CHAIR
- STORAGE UNIT 24" × 39"
- FOLDDOWN BED 66" × 33"
- BATHTUB 66" × 30"
- BATHTUB 60" × 30"
- CORNER BATHTUB 48" SQ
- SHOWERS 48" SQ, 36" SQ, 30" SQ
- WHIRLPOOL 48" SQ
- BIDET 12" × 20"
- LAVATORIES 20" × 20", 20" × 20", 24" × 20"
- POTTY CHAIR 14" SQ
- WATER CLOSET 24" × 30"
- COUNTER TOP 18" × VARIES
- URINAL 18" × 12"
- LAUNDRY HAMPER 18" SQ
- LAZY SUSAN / FLOOR CAB 24"
- FREEZER 36" × 24"
- SNACK COUNTER 18" × VARIES
- STOOLS 10" DIA
- FREEZER REFRIGERATOR 36" × 26"
- STOVE 34" × 25"
- STOVE 22" × 25"
- KITCHENETTE 26" × 50"
- WALL CABINET 12"
- FLOOR CABINET 24" × VARIES
- REF 27" × 24"
- REFRIG 30" × 24"
- SINKS 12" × 18", 16" × 18", 24" × 18"
- OVEN 24" × 24"
- LAUNDRY TUBS 22" × 22", 26" × 26"
- WASHER 30" × 28"
- DRYER 30" × 28"
- TRASH COMPACTOR 20" × 22"

T-4: LIVING AND SLEEPING AREAS—ELEVATIONS

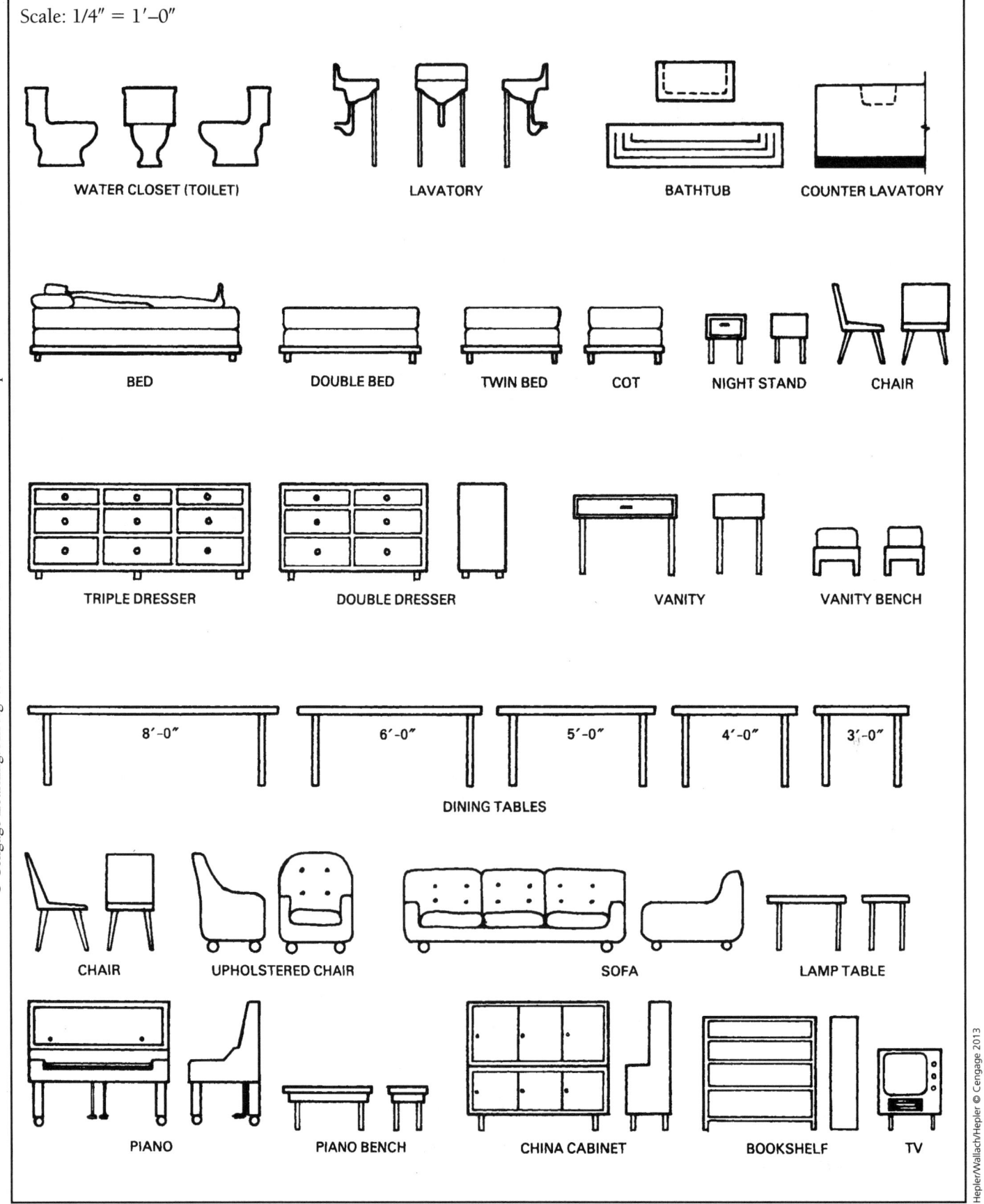

Scale: 1/4" = 1'–0"

WATER CLOSET (TOILET) LAVATORY BATHTUB COUNTER LAVATORY

BED DOUBLE BED TWIN BED COT NIGHT STAND CHAIR

TRIPLE DRESSER DOUBLE DRESSER VANITY VANITY BENCH

DINING TABLES: 8'-0" 6'-0" 5'-0" 4'-0" 3'-0"

CHAIR UPHOLSTERED CHAIR SOFA LAMP TABLE

PIANO PIANO BENCH CHINA CABINET BOOKSHELF TV

T-5: SERVICE AREA—ELEVATIONS

Scale: 1/4" = 1'-0"

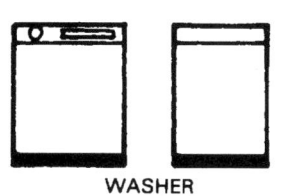

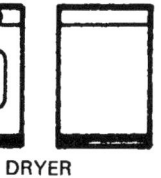

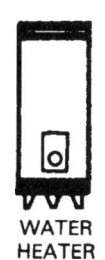

WASHER

DRYER

WATER HEATER

FREEZER

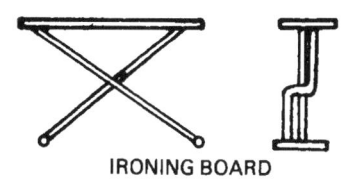

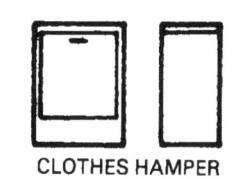

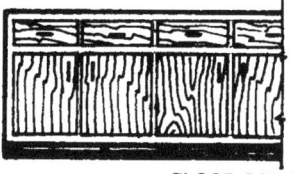

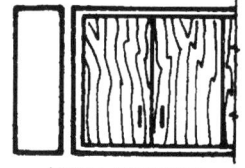

LAUNDRY TRAY

IRONING BOARD

CLOTHES HAMPER

FLOOR CABINET

WALL CABINET

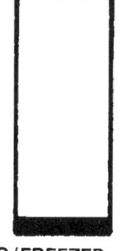

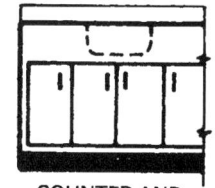

RANGES

REFRIGERATOR/FREEZER

DISHWASHER

COUNTER AND SINK

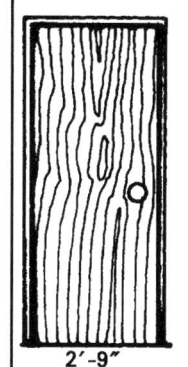

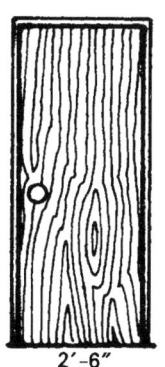

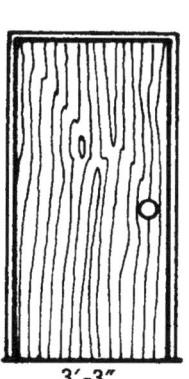

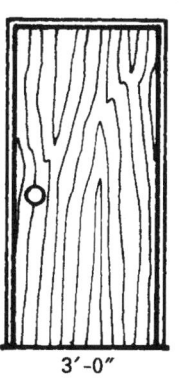

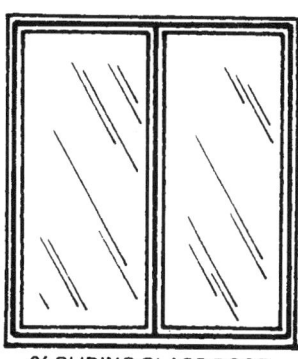

2'-9" 2'-6" 3'-3" 3'-0"

DOORS

6' SLIDING GLASS DOOR

PEOPLE

2'-6" × 2'-0"

WINDOWS

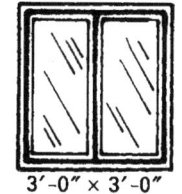

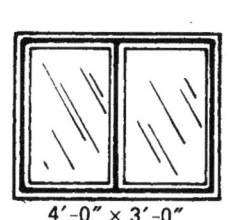

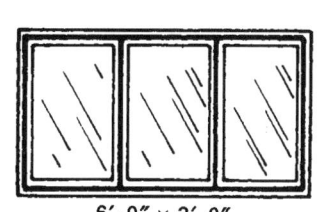

3'-0" × 3'-0"

4'-0" × 3'-0"

6'-0" × 3'-0"

SECTION 3
Chapter Review Tests

The following pages contain tests that relate to each chapter in the student text. The answer key for these tests is located in the Instructors Guide. The answers are based on the most commonly used terms; however, there are synonyms for the same architectural terms used in different geographical areas and by different contractors. Students should use the form presented by the instructor. In selecting an answer to which a leader points, always select the smallest component, not an entire assembly.

Chapter 1 Review Test

Directions: In the space at the left, write the letter of the choice that best completes each statement.

____ 1. Solid walls that support themselves are called
    a. bearing walls
    b. post-and-lintels
    c. vaults
    d. skeleton frames

____ 2. An open frame wall to which a covering is attached is called a
    a. bearing wall
    b. post-and-lintel
    c. cantilever
    d. skeleton frame

____ 3. Historically, the first solution to the problem of making an opening in a supporting wall without sacrificing the support was the
    a. bearing wall
    b. post-and-lintel
    c. skeleton frame
    d. flying buttress

____ 4. Many arches arranged in a circle form a
    a. keystone
    b. vault
    c. dome
    d. cantilever

____ 5. The supporting stone in an arch is called a
    a. vault
    b. barrel vault
    c. keystone
    d. cross support

Directions: Match each item or description in the left-hand column with the correct item or description in the right-hand column. Write the letter of your choice in the space provided.

____ 6. Half-timbered with mortar between, adopted from Tudor style
____ 7. Norman towers and tall chimneys
____ 8. Similar to Spanish architecture
____ 9. Mansard roof
____ 10. Courtyard patio
____ 11. Gambrel roof
____ 12. Freedom of design and use of space
____ 13. Rambling plan
____ 14. Intricate finials, lintels, parapets, balconies, and cornices
____ 15. Cape Cod style
____ 16. Two-story columns

a. Dutch colonial
b. Elizabethan
c. English Tudor
d. French provincial
e. Italian
f. Modern
g. New England style
h. Ranch style
i. Southern colonial
j. Spanish
k. Victorian

Directions: Write your answer to each question in the space provided.

17. What is the term for a double-pitched roof?
18. Name four architectural styles developed in colonial America.
19. The development of what material made large, high-rise buildings possible?
20. What is the name for a protruding structure used to add support to an arch or wall?
21. Name the four European styles that have had the most influence on American architecture.
22. What is another name for Mid-Atlantic style architecture?

Chapter 2 Review Test

Directions: In the space at the left, write the letter of the choice that best completes each statement.

_____ 1. Form follows
   a. shape
   b. space
   c. function
   d. trends

_____ 2. Which design term describes the quality of being useful?
   a. creativity
   b. functionalism
   c. aesthetic value
   d. trends

_____ 3. Which of the following is *not* one of the seven basic elements of design?
   a. proportion
   b. line
   c. texture
   d. color

_____ 4. What are groups of colors that relate well to each other called?
   a. triadics
   b. harmonies
   c. shades
   d. chromas

_____ 5. Which principle of design is used to draw attention to a certain area?
   a. emphasis
   b. unity
   c. balance
   d. transition

_____ 6. Which term is used to describe the center of attention in a design?
   a. neutral point
   b. emphasis
   c. form
   d. focal point

Directions: Write your answer to each question in the space provided.

_____ 7. Name the three primary colors.

_____ 8. What is the term for a color created by combining two primary colors?

_____ 9. What term is used to describe the lightness or darkness of a color?

_____ 10. What is the term for colors opposite one another on the color wheel?

_____ 11. When two halves of a design are almost the same, which type of balance is created?

_____ 12. What is the term for a pleasant change from one element to another without destroying the unity of the design?

_____ 13. Combining a primary color with a neighboring secondary color produces which type of color?

Directions: Match each item or description in the left-hand column with the correct item or description in the right-hand column. Write the letter of your choice in the space provided.

____ 14. Same pattern used over and over; creates a sense of motion

____ 15. Reflect from surfaces

____ 16. Roughness

____ 17. Assures appropriate sizing of components

____ 18. Equilibrium

____ 19. Surrounds form

____ 20. Use of contrasting elements for variety

____ 21. Can create a sense of height

____ 22. Creates a sense of flowing movement

____ 23. Creates a sense of wholeness

____ 24. Produced by adding black to a color

____ 25. Another term for shape

____ 26. White, gray, or black

a. Proportion
b. Shade
c. Rhythm
d. Texture
e. Neutral
f. Curved line
g. Unity
h. Line
i. Form
j. Light and shadow
k. Space
l. Balance
m. Opposition

Chapter 3 Review Test

Directions: Write your answer to each question in the space provided.

_____ 1. Name the three types of scales used to prepare architectural drawings.

_____

_____

_____ 2. Which type of scale has all increments shown along its length?

_____ 3. To what scale are residential floor plans and elevations most often drawn?

_____ 4. What are the two basic metric units of measure used on architectural drawings?

_____ 5. What type scale is used in drawing survey plans?

_____ 6. What type scale has only the main units marked along the length of the scale?

_____ 7. What type of scale is used to measure a dimension of 3'-7&1/8"?

_____ 8. What type scale is used to measure a dimension of 9.3"?

_____ 9. On an architects scale if ½" equals 1" what does ¼" represent?

_____ 10. A scale of 1" = 1' − 0" decreases the actual drawing by how many times?

_____ 11. What type scale uses ratios in increments of 10?

_____ 12. What type scale is divided into fractional units?

Chapter 4 Review Test

Directions: In the space at the left, write the letter of the choice that best completes each statement.

____ 1. Which type of lines permits dimensioning at a distance from a structure?
  a. extension
  b. hidden
  c. phantom
  d. cutting plane

____ 2. A rendering does not include
  a. lines
  b. shadows
  c. dimensions
  d. landscaping

____ 3. What is used to reference a door or window on a plan to a door or window schedule?
  a. border
  b. callout
  c. leader line
  d. template

____ 4. Which type of drawing gives a "bird's eye" view?
  a. section
  b. detail
  c. plan
  d. elevation

____ 5. Drawings used during the building process are called
  a. working drawings
  b. details
  c. general-purpose drawings
  d. building drawings

____ 6. Under the coding system, which letter is used to identify a set of interior design drawings?
  a. I
  b. X
  c. D
  d. A

Directions: Write your answer to each question in the space provided.

_____ 7. What very heavy lines are used to show where an area is to be sectioned?

_____ 8. What lines are used to connect a note or dimension to a drawing?

_____ 9. What coding system is used to organize and mark architectural drawings consistently?

_____ 10. How is each drawing sheet identified on its bottom and/or right side?

_____ 11. What is another term for a pictorial drawing?

_____ 12. What is the term for a 3D replica of a structure?

_____ 13. Which type of drawing is done to reveal precise information about construction methods or materials?

_____ 14. Which type of line is used to indicate alternate positions of moving parts?

194 ■ Student Workbook to Accompany Drafting and Design for Architecture and Construction

_____ 15. Fractions are drawn how many times the height of a whole number?

_____ 16. Which type of drawing is this?

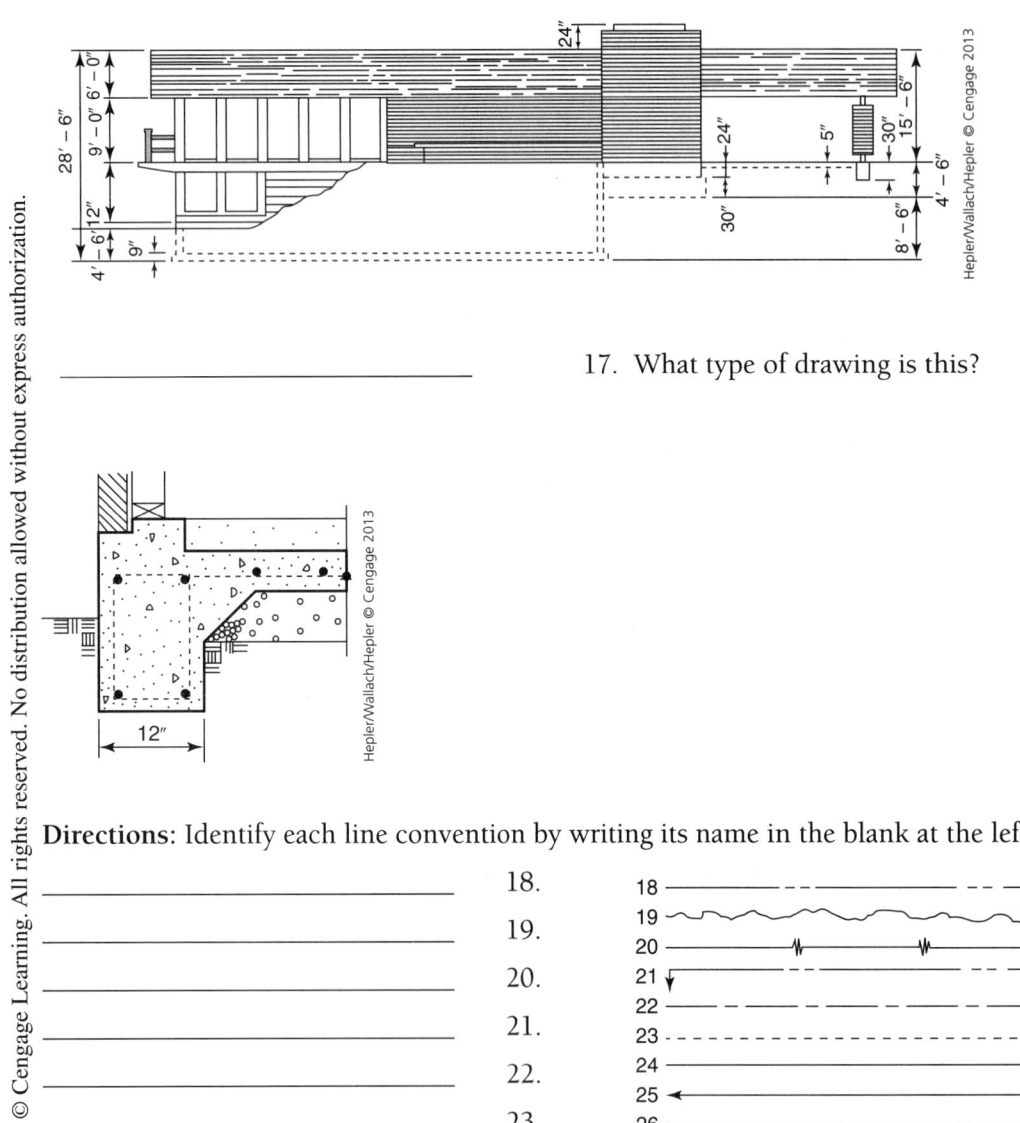

_____ 17. What type of drawing is this?

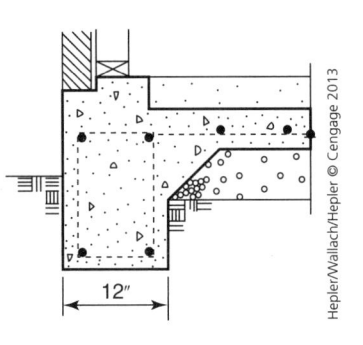

Directions: Identify each line convention by writing its name in the blank at the left.

_____ 18.

_____ 19.

_____ 20.

_____ 21.

_____ 22.

_____ 23.

_____ 24.

_____ 25.

_____ 26.

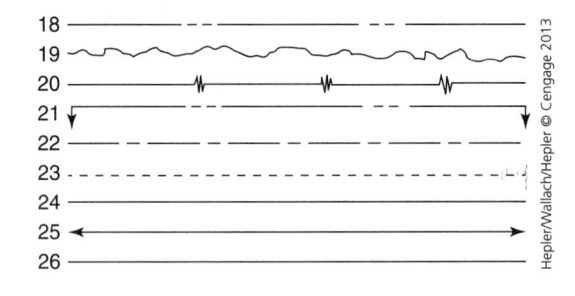

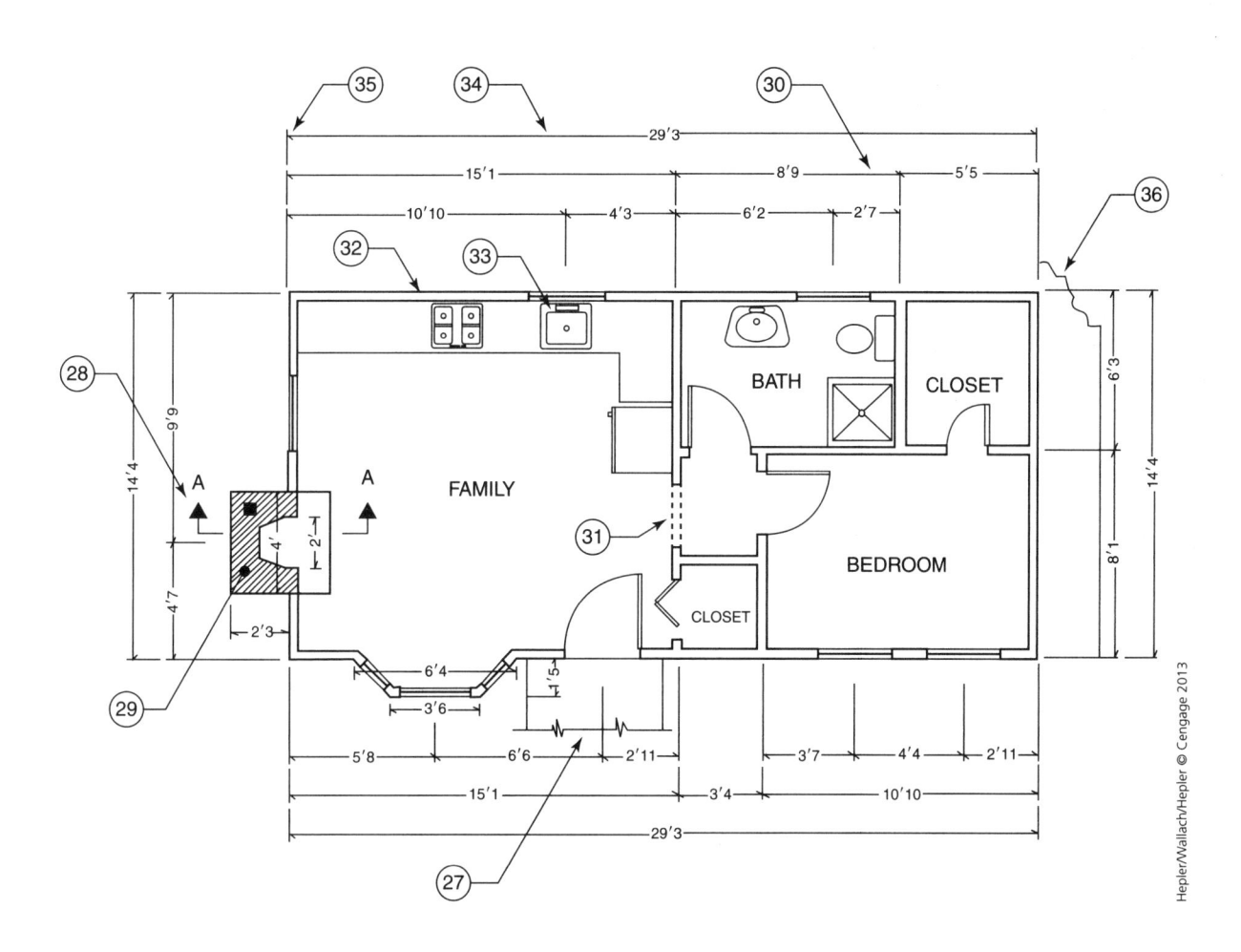

Chapter 4 Review Test ■ 195

_____ 27.
_____ 28.
_____ 29.
_____ 30.
_____ 31.

_____ 32.
_____ 33.
_____ 34.
_____ 35.
_____ 36.

Chapter 5 Review Test

Directions: Write your answer to each question in the space provided.

_____ 1. What hardware component is considered the "brain" of a computer system?

_____ 2. What computer hardware device resembles a TV screen?

_____ 3. What is the term for a computer's temporary memory, which determines how much software can run simultaneously on the computer?

_____ 4. What term is used for an arrow that can be moved around on a monitor screen?

_____ 5. What is the hardware component that controls the pointer?

_____ 6. What term is used to describe the crosshair input device used to pick points on a tablet?

_____ 7. What is the name of a single point of light on a monitor?

_____ 8. What device uses pens or pencils of various kinds to create a printout of a large architectural drawing?

_____ 9. What device is used to link a computer system with a telecommunications network?

_____ 10. What is the term for computers that are capable of combining text, graphics, audio, animation, and full-motion video?

_____ 11. What term is used to describe the individual lines, circles, etc., created on a CAD system?

_____ 12. Which command allows an object to be turned around a given point?

_____ 13. Which command is most commonly used in preparing architectural drawings?

_____ 14. Which command is used to duplicate an object and place it elsewhere on a drawing?

_____ 15. Which command is used to place an object in another location on a drawing?

_____ 16. Which command magnifies or reduces the apparent size of a drawing on the screen?

_____ 17. Which command is used to move the view window (part of the drawing that is visible) horizontally or vertically on a screen?

_____ 18. Name the Cartesian coordinate axes used to create a three-dimensional drawing.

_____ 19. What term describes 3D drawings in which all lines, intersections, and hidden areas are exposed?

_____ 20. What term describes 3D drawings that consist of solid plane surfaces?

_____ 21. What term describes 3D drawings having mass properties that can be measured and rendered?

_____ 22. What term describes 3D technology that allows people to "move" freely through a building to view it from the inside?

Directions: Place the letter representing the correct answer on the left column line.

____ 23. What does the acronym CAD stand for?
   a. computer-aided drafting
   b. computing with active designs
   c. computer-aided design
   d. contemporary architectural design

____ 24. The resolution of a CAD system's monitor depends on the
   a. size of the monitor's screen
   b. multicolored screen
   c. monochrome screen
   d. number of pixels
   e. all of the above

____ 25. What holds the CAD system's temporary memory?
   a. ROM
   b. RAM
   c. CAD
   d. pixels

____ 26. What does the abbreviation CPU stand for?
   a. central processing unit
   b. circuit processing unit
   c. command principle user
   d. Cartesian presentation user

____ 27. CAD inputs may be performed with a(n)
   a. mouse
   b. alphanumeric keyboard
   c. voice command
   d. pressure plate
   e. all of the above

Chapter 6 Review Test

Directions: Write your answer to each question in the space provided.

_____ 1. Which type of solar design system uses mechanical or electrical devices to control the energy?

_____ 2. In the northern hemisphere, which two sides of a building receive the most sunlight?

_____ 3. On which side of a building would you locate a patio in order to receive maximum shade?

_____ 4. Name the three basic areas of a lot.
_____
_____
_____

_____ 5. The exposed wall of an earth-sheltered house should face in what compass direction?

_____ 6. Which type of trees should be planted to maximize summer cooling and warmth from the sun in winter?

_____ 7. Which type of trees are most effective in blocking wind?

_____ 8. What always replaces rising warm air?

_____ 9. On southern slopes, wind patterns move in which direction during the day?

_____ 10. What is another term for the wind tunnel effect?

_____ 11. What is the name of the science of designing and arranging the things people use?

_____ 12. Name the five areas of potential pollution that must be addressed during the architectural design process.
_____
_____
_____
_____

_____ 13. What is the unit of sound measurement?

_____ 14. What is the term for positioning structures in relation to environmental conditions?

_____ 15. Which side of a house (compass direction) is the coolest?

_____ 16. Which side of a house provides a view of the sunset?

_____ 17. Which side of a house is brightest and coolest in the morning?

_____ 18. Which side of a house is fully shaded?

_____ 19. Which side of a house needs maximum overhang protection?

_____ 20. Which side of a house will retain the most late afternoon heat?

Directions: In the space at the left, write the letter of the choice that best completes each statement.

_____ 21. Building sites are sold and registered as
   a. lots
   b. plots
   c. soil
   d. public areas

_____ 22. The major concern for earth-sheltered homes is
   a. shelter
   b. heat loss
   c. waterproofing
   d. vegetation

_____ 23. Indoor heat can be lost through crevices around doors and windows by means of
   a. landscaping
   b. wind chill
   c. baffling
   d. orientation

_____ 24. The removal of healthy trees from large tracts of land is a form of
   a. ecology
   b. orientation
   c. land pollution
   d. ergonomics

Directions: Select the correct answer and write the letter for it in the space at the right.

_____ 25. Which of the following is a major characteristic of passive solar orientation?
   a. There are solar collectors with pumps and pipes
   b. It will function only in the summer
   c. It has no moving mechanical parts
   d. There is abundant insulation to keep heat out of the house   _____

_____ 26. Which of the following is a major characteristic of an active solar heating system?
   a. There are solar collectors with pumps and pipes
   b. It will function only in the summer
   c. It has no moving mechanical parts
   d. There is abundant insulation to keep heat out of the house   _____

_____ 27. When soil is placed along the side of a structure to help stop heat transfer, it is called a
   a. dirt insulator   b. frem   c. soil conditioner   d. berm   _____

_____ 28. Which part of a home will be the warmest in the early evening?
   a. northeast   b. southeast   c. southwest   d. northwest   _____

_____ 29. What single major factor has caused society to become energy conscious?
   a. More jobs are provided in the building industry
   b. The cost of energy is increasing
   c. There is goodwill toward the oil-producing countries
   d. Our oil supply may someday be depleted   _____

_____ 30. In the northern hemisphere, which side of a home receives the most sun?
   a. north   b. south   c. east   d. west   _____

_____ 31. Which type of landscaping is the most efficient for solar orientation?
   a. deciduous trees
   b. large evergreen trees
   c. large areas of low shrubbery
   d. large expanses of grass   _____

_____ 32. Which material is *not* a good heat insulator?
   a. air space   b. vacuum space   c. single glazing
   d. wood   e. double glazing   f. styrofoam   _____

_____ 33. On which side of a home should a greenhouse be located for efficient passive solar orientation?
   a. north   b. south   c. east   d. west

Chapter 7 Review Test

Directions: Write your answer to each question in the space provided.

_____ 1. What is the term for an open area that combines the functions of a living room, dining room, and family room?

_____ 2. In which type of plan are living area rooms separated by solid partitions and doors?

_____ 3. What is the term used to describe an indoor garden room?

_____ 4. What is another term for "pattern of decoration"?

_____ 5. What living area room must be adjacent to the kitchen?

_____ 6. Which living area room should have direct access to both the dining room and the entry?

_____ 7. What is another term for the activities or multiactivities room?

_____ 8. What are the two main purposes of a living area window?

_____
_____
_____ 9. Name the three types of lighting used in a living area.

_____
_____
_____ 10. What devices can be used with a floor plan to plan the placement of furniture before room sizes are determined?

_____ 11. What is the term for a dining room wall having an opening for passing food through from the kitchen?

_____ 12. What is another term for a dimmer switch?

_____ 13. What is the minimum space that should be provided between a dining chair and a wall or furniture?

_____ 14. What is the recommended distance between dining chair centers?

Directions: In the space at the left, write the letter of the choice that best completes each statement.

_____ 15. Which of the following is *not* important to a living room?
   a. nearness to the kitchen
   b. the most attractive view
   c. central location
   d. nearness to the entry

_____ 16. Which of the following is *not* a use for a fireplace?
   a. decoration
   b. heat
   c. ventilation
   d. partition

_____ 17. Most ceilings are covered with plaster or
   a. plastic laminates
   b. gypsum board
   c. hardwood
   d. masonry

_____ 18. A separate dining room should seat _____ people.
   a. 6–12
   b. 10–15
   c. 8–9
   d. 4–6

_____ 19. Recreation rooms are often located
   a. upstairs
   b. near quiet areas
   c. near a laundry area
   d. in the basement

_____ 20. Which is *not* an example of a special-purpose room?
   a. office
   b. den
   c. game room
   d. exercise room

Chapter 8 Review Test

Directions: Write your answer to each question in the space provided.

_____

1. Which two architectural styles are characterized by multiple balconies?

2. Name the three basic types of patios according to function.

3. Name the three main factors that affect pool location.

4. How many square feet are covered by a pool that measures 14′ × 24′?

5. Name the three materials normally used to frame swimming pools.

6. Name two finishes used on underwater pool surfaces.

7. What is the minimum water depth at the end of the pool where the diving board is located?

8. How many feet must a diving well extend horizontally from the end of the board?

9. What is the minimum fence height for pool enclosures required by most building codes?

10. How many cubic feet are contained in a pool that measures 12.5′ × 22′ × 6′?

11. Name the two pool safety devices normally required by building codes.

12. What is the mathematical formula used to calculate the area of a round pool?

Directions: Match each item or description in the left-hand column with the correct item or description in the right-hand column. Write the letter of your choice in the space provided.

___ 13. Patio
___ 14. Breezeway
___ 15. Loggia
___ 16. Porch
___ 17. Living patio
___ 18. Balcony or deck
___ 19. Lanai
___ 20. Spanish patio
___ 21. Veranda
___ 22. Quiet patio

a. Exterior covered passageway
b. Covered platform leading to an entrance
c. Wrap-around porch
d. Courtyard
e. Master bedroom patio
f. Patio adjacent to great room
g. Italian roofed arcade
h. Roofed area between buildings
i. Ground-level exterior living surface
j. Elevated porch

Directions: Place the letter representing the correct answer on the left column line.

___ 23. The outdoor living areas may include
   a. decks
   b. patios
   c. porches
   d. pools
   e. all of the above

___ 24. Which indoor area is well suited for a patio?
   a. entry
   b. sleeping
   c. living
   d. service
   e. all of the above

___ 25. What is the area of a pool that is 20′ × 10′ and 5′ deep?
   a. 100 square feet
   b. 200 square feet
   c. 300 square feet
   d. 500 square feet

___ 26. What is the volume of the pool in question 3?
   a. 1,000 cubic feet
   b. 2,000 cubic feet
   c. 3,000 cubic feet
   d. 4,000 cubic feet

___ 27. What is the area of a round garden pool with a 10-foot radius?
   a. 114 square feet
   b. 214 square feet
   c. 314 square feet
   d. 414 square feet

Chapter 9 Review Test

Directions: Write your answer to each question in the space provided.

_____ 1. What is the term for the "step" part of a stair?

_____ 2. What is the term for the vertical part of a stair?

_____ 3. Which type of switch should be used to control stair lighting?

_____ 4. What is the average residential riser height?

_____ 5. What is the minimum stair and hall width commonly required by building codes?

_____ 6. What is the term for a platform used to change stair direction?

_____ 7. What is the term for the area opening used for stairs?

_____ 8. What is the term for the vertical distance between the top of each stair tread and the ceiling?

_____ 9. Name the four basic types of entrances.

_____

_____

_____

_____ 10. What is the term for an inside entrance area?

_____ 11. What is the term for a driveway extension used to park or turn automobiles?

_____ 12. What is the minimum stairwell headroom required by most building codes?

_____ 13. What is the minimum driveway width required by most building codes?

_____ 14. What is the minimum inside turning radius required for driveways?

_____ 15. A landing should be planned for stairs having more than how many risers?

Chapter 11 Review Test

Directions: In the space at the left, write the letter of the choice that best completes each statement.

_____ 1. Which of the following is *not* usually designed into a laundry area?
   a. sewing area
   b. ironing area
   c. service entrance
   d. half-bath

_____ 2. Standard garage doors range in height from
   a. 6'-6" to 8'-0"
   b. 6'-5" to 9'-0"
   c. 10'-0" to 12'-0"
   d. 8'6" to 12'-6"

_____ 3. Standard garage doors range in width from
   a. 7'-0" to 9'-0"
   b. 12'-0" to 18'-0"
   c. 7'-0" to 20'-0"
   d. 9'-6" to 17'-6"

_____ 4. Which paint finish should be used for workshops?
   a. high gloss
   b. semi-gloss
   c. matte
   d. any of the above

_____ 5. What should be added to a laundry if excess moisture is anticipated?
   a. window
   b. electric dryer
   c. dehumidifier
   d. exhaust fan

Directions: Write your answer to each question in the space provided.

_____ 6. Which type of voltage outlets should be provided in the laundry?

_____ 7. How far above a laundry appliance should a light fixture be placed?

_____ 8. Name the three basic types of garages.

_____ 9. What is the minimum slab thickness for most residential garages?

_____ 10. Name the four common types of garage doors.

_____ 11. Name four materials usually used to manufacture garage doors.

_____ 12. What three materials are most commonly used to surface driveways?

_____ 13. What is used to prevent concrete driveways from cracking?

14. Name three basic types of closets.

15. What is the minimum depth of a wardrobe closet?

16. Depths of more than how many inches make reaching the back of a closet difficult?

17. Name four materials used for closet shelving.

18. What is the recommended height for a workbench?

19. What two items can be used in wall and ceiling construction to reduce the amount of noise passing to other rooms from a workshop?

20. Name the two types of workbench.

Chapter 12 Review Test

Directions: Write your answer to each question in the space provided.

_____ 1. What can be used to provide ventilation for an interior bath?

_____ 2. Which type of bath is divided by partitions?

_____ 3. In which type of bath are all fixtures visible?

_____ 4. What is the minimum clearance between the center of a water closet and a side wall or other fixture?

_____ 5. What is the minimum size of a bathroom containing all three basic fixtures?

_____ 6. Which three fixtures must be included in any full bath?

_____ 7. What is the size range for an average-sized bedroom in square feet?

_____ 8. Bedroom closets should include at least how many feet of clothes-hanging rod per person?

_____ 9. Which type of windows can help reduce noise in a bedroom?

_____ 10. If there is no air conditioning, which type of ventilation should be designed into a bedroom plan?

_____ 11. Which three furniture items must a minimum-size bedroom accommodate?

_____ 12. How wide must a bedroom door be to allow passage of furniture?

_____ 13. Which type of window provides for privacy while allowing furniture to be placed underneath?

_____ 14. Name at least two materials used in the manufacture of lavatories or sinks.

_____ 15. Name the three types of lavatories used in bathrooms.

_____ 16. What is a comfortable sink height for most people?

_____ 17. Name the two basic types of water closets.

_____ 18. What is the minimum wheelchair clearance needed between the water closet center and a side wall or projection?

_____ 19. Name three qualities that materials used in bath decor should have.

_____

_____

_____ 20. Which two bathroom fixtures are included in a half-bath?

_____

Directions: Place the letter representing the correct answer on the left column line.

_____ 21. Approximately which percentage of time in a day does the average adult spend sleeping?
   a. 20%   c. 50%
   b. 30%   d. 100%

_____ 22. What is the minimum width of a wardrobe closet?
   a. 15"   c. 24"
   b. 20"   d. 30"

_____ 23. What is the minimum width of a two-rod walk-in closet?
   a. 6'   c. 8'
   b. 4'   d. 10'

_____ 24. What is the minimum depth of a one-rod walk-in closet?
   a. 8'   c. 5'
   b. 6'   d. 4'

_____ 25. What is the minimum clearance for the front of a toilet?
   a. 36"   c. 18"
   b. 24"   d. 15"

_____ 26. What is the minimum space clearance for the placement of a toilet?
   a. 18"   c. 36"
   b. 24"   d. 42"

Chapter 13 Review Test

Directions: Write your answer to each question in the space provided.

_____ 1. Which laws regulate the uses of a site in order to preserve public safety?

_____ 2. Which building code limit allows more light to reach adjacent properties on the north side?

_____ 3. Name the three major zones into which most municipalities are divided.

_____

_____

_____ 4. What is the term used for the percentage of land in a community on which buildings cannot be built?

_____ 5. What is the term used for the space a building takes up on its lot?

_____ 6. What is the term for the ratio of all inhabitants to space in a specific geographic area?

_____ 7. What document must be obtained from a local building department before building can begin?

_____ 8. Which drawing shows the exact size, shape, and levels of a site?

_____ 9. Which type of lines connect points on a site that are at the same elevation?

_____ 10. What is the term for a survey of multiple connected properties?

_____ 11. Which kind of plan shows the size and shape of a site and the location and size of all buildings?

Directions: Match the descriptions with the map elements by writing the correct letters in the spaces provided. Some items have more than one answer.

_____ 12. Stream

_____ 13. Property corner elevation symbol

_____ 14. Contour line

_____ 15. Stream bank elevation

_____ 16. Compass direction of property line

_____ 17. Gas line (4")

_____ 18. Lot owner

_____ 19. Compass direction

_____ 20. Property line

_____ 21. Elevation of tree

_____ 22. Elevation of manhole cover

_____ 23. Center of street

_____ 24. Lot identification

_____ 25. Manhole cover

_____ 26. Tree location

_____ 27. Property corner elevation

_____ 28. Property dimension

_____ 29. Stream bank

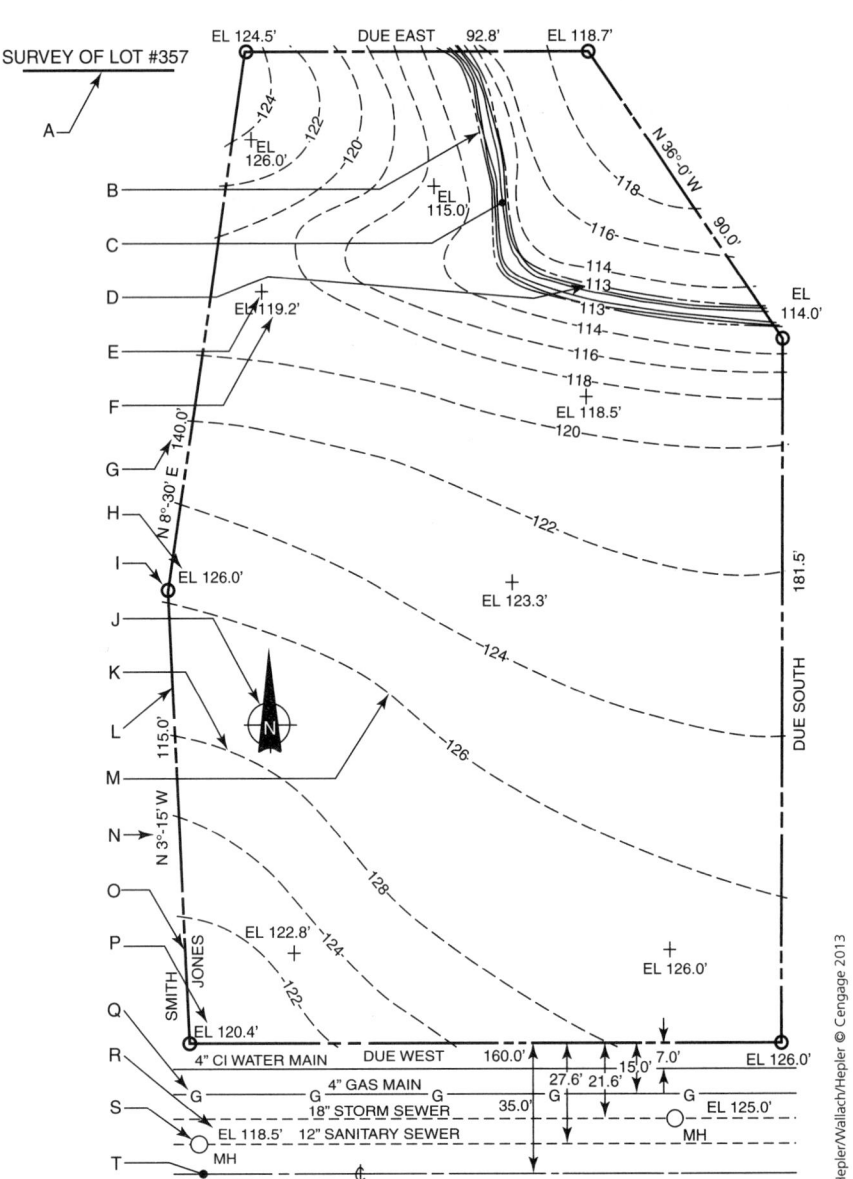

Chapter 16 Review Test

Directions: Write your answer to each question in the space provided.

_____ 1. Which measurements are found on elevation drawings that cannot be shown on floor plans?

_____ 2. Name the three main horizontal lines shown on elevation drawings.
_____
_____

_____ 3. Which elevation feature protects walls and windows from sun and precipitation?

_____ 4. What is the name for the horizontal distance between the roof ridge and an outside wall?

_____ 5. In calculating roof pitch, what is the term for the ratio of the rise over the run?

_____ 6. What is the term for the vertical distance between the top of a wall supporting a roof and the ridge?

_____ 7. What is the term used to describe the arrangement of windows in an elevation?

_____ 8. The outline of which roof type follows the triangular end of the building?

_____ 9. Which roof type provides eave-line protection around the entire perimeter of the building?

_____ 10. Which roof type, if facing south, is suited to the use of solar panels?

_____ 11. What is the term for the underside of a roof overhang?

Directions: Write the name of each roof style in the space provided.

_____ 12.
_____ 13.
_____ 14.
_____ 15.
_____ 16.
_____ 17.

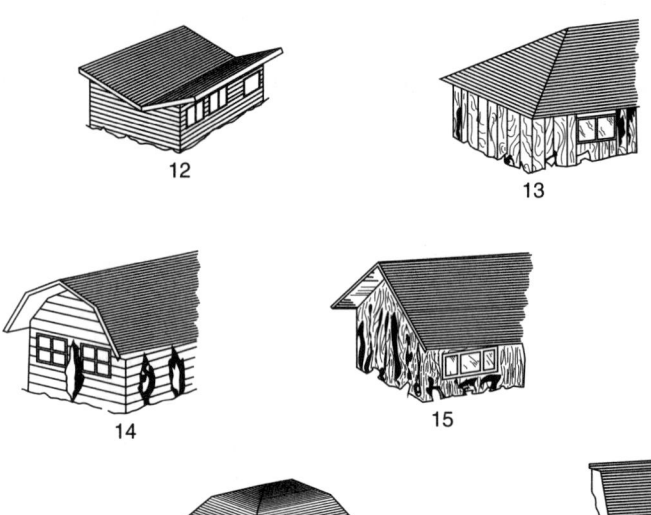

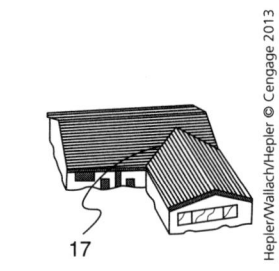

Directions: Identify the window styles by writing the correct letters in the spaces provided.

____ 18. Awning
____ 19. Casement
____ 20. Double-hung
____ 21. Hopper
____ 22. Jalousie
____ 23. Sliding

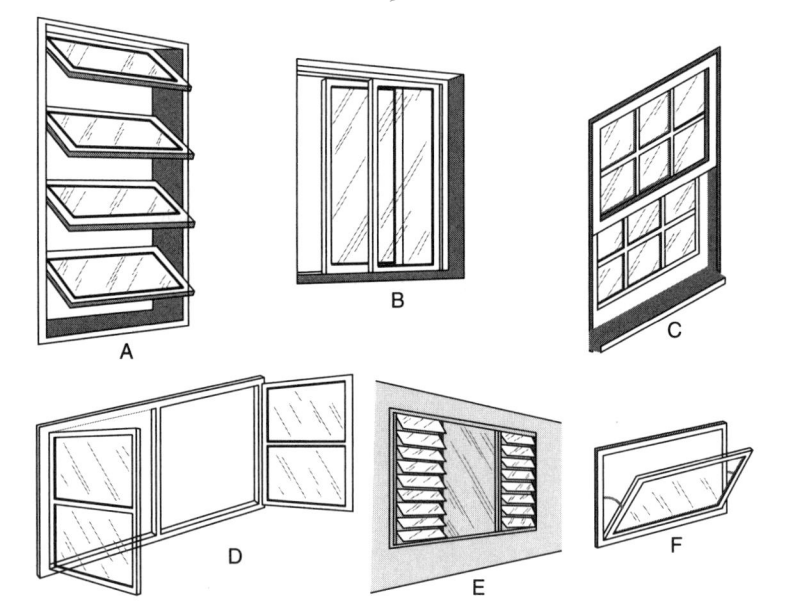

Chapter 17 Review Test

Directions: Write your answer to each question in the space provided.

_____ 1. What is another term for multiview projection?

_____ 2. Name the six views in a multiview drawing.

_____

_____

_____

_____

_____

_____ 3. If the front view of a structure is the south elevation, in which compass direction does the right elevation face?

_____ 4. If the left view of a structure is the east elevation, in which compass direction does the rear elevation face?

_____ 5. Exterior elevations are always drawn to the same scale as which other drawing?

_____ 6. Which type of line is used to draw major features below the ground line on an exterior elevation?

_____ 7. What term is used to describe the elevation drawing of a site?

_____ 8. When a wall on a floor plan is not drawn at a 90° angle, which type of elevation must be done?

_____ 9. What is the term used for the distance between roof supports?

_____ 10. Which constant numerical unit is assigned to the run when calculating roof pitch?

_____ 11. What is the name for a drawing of an inside wall?

_____ 12. On an elevation window drawing, what does the point of a dashed line represent?

_____ 13. What is used to key a window or door to a door or window schedule?

_____ 14. What is the term for the angle of a roof described as a ratio?

_____ 15. Which dimension line on an elevation drawing remains constant?

_____ 16. Vertical dimensions should read from which side of a drawing?

_____ 17. When calculating pitch, how do you find the span?

_____ 18. The run of a certain roof is 12. The rise is 6. What is the pitch?

_____ 19. Which parts of a window should appear on all elevation drawings?

_____ 20. What is the term used for an elevation drawing to which landscape and material textures have been added?

Directions: Place the letter representing the correct answer on the left column line.

____ 21. What is the term used for the projections of exterior elevation drawings?
    a. parallel
    b. perpendicular
    c. in-line
    d. orthographic

____ 22. What angle of projection, from the floor plan, is not necessary for an auxiliary elevation?
    a. 30°
    b. 60°
    c. 90°
    d. 80°

____ 23. Which style of roof has only two sloping surfaces?
    a. gable
    b. hip
    c. flat
    d. gambrel

____ 24. What type of window is hinged at its sides?
    a. slider
    b. casement
    c. double hung
    d. swing

____ 25. What is the standard residential ceiling height for interior elevations?
    a. 7'
    b. 8'
    c. 9'
    d. 10'

Chapter 18 Review Test

Directions: Write your answer to each question in the space provided.

_____ 1. Name the two main types of sectional drawings.

_____

_____ 2. Why is a cutting plane line drawn on a floor plan?

_____ 3. To which scale are full sections often drawn?

_____ 4. Which type of line is used on a drawing to indicate that a portion of the object has been removed?

_____ 5. Name at least two areas of a wall section that are commonly drawn as removed sections.

_____

_____ 6. Internal construction of door components is shown with which type of drawing?

_____ 7. Which shortcut method is used to dimension cabinets?

Directions: In the space at the left, write the letter of the choice that best completes each statement.

_____ 8. Which type of drawing is created using a cutting plane along a building's length?
    a. offset section   c. removed section
    b. longitudinal section   d. full section

_____ 9. Sill sections show how the foundation intersects with the floor system and the
    a. outside wall   c. gutter
    b. cornice   d. ridge

_____ 10. A footing section shows the position of the
    a. cornice   c. foundation wall
    b. beams and columns   d. all of the above

_____ 11. A base section shows how the wall-finishing materials are attached to the
    a. foundation   c. crown
    b. beams   d. studs

_____ 12. When a cutting plane goes across the entire window, the result is a
    a. crown section   c. base section
    b. sill section   d. jamb section

Directions: Match the terms to the pictured items by writing the correct letters in the spaces provided.

____ 13. Brick

____ 14. Center line

____ 15. Vertical siding

____ 16. Run

____ 17. Rise

____ 18. Roof covering

____ 19. Fascia or barge rafter

____ 20. Overhang

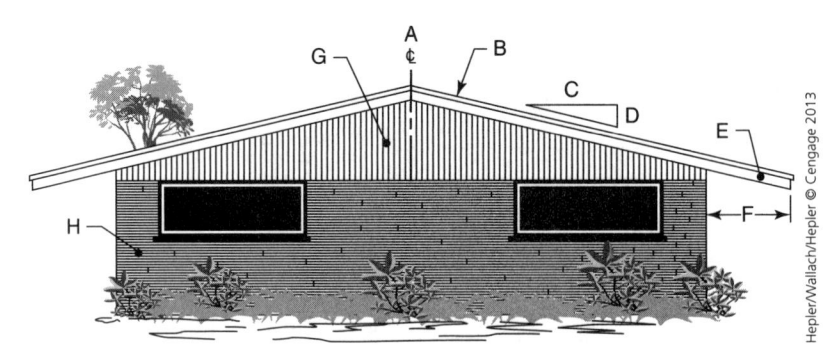

Chapter 19 Review Test

Directions: Write the answer to each question in the space provided.

_____ 1. Name the three basic types of pictorial drawings.
_____
_____

_____ 2. Which type of drawing is created by adding an angled side to an elevation drawing?

_____ 3. Which type of drawing includes receding lines drawn at 30° from the horizon?

_____ 4. What is the term for the spot at which the receding lines on a perspective drawing seem to disappear?

_____ 5. Which line on a perspective drawing represents the observer's eye level?

_____ 6. What is the term for the exact location of the observer on a perspective drawing?

_____ 7. Which is the only line shown in its true length on a two-point perspective drawing?

_____ 8. Which type of pictorial drawing is best for drawing construction details?

_____ 9. In an interior one-point perspective drawing, where should the vanishing point be placed to show equal amounts of both side walls, floor, and ceiling?

_____ 10. Which type of drawing is used to overcome the height distortion of tall buildings?

_____ 11. What imaginary surface stands between the station point and the object being drawn in a perspective drawing?

Directions: Identify each type of drawing by writing its description in the space provided.

_____ 12.
_____ 13.
_____ 14.
_____ 15.
_____ 16.
_____ 17.
_____ 18.

Directions: Place the letter representing the correct answer on the left column line.

_____ 19. On which type of perspective drawing may one surface be drawn to scale?
    a. one-point   c. three-point
    b. two-point   d. none of the above

_____ 20. At what angle are the parallel receding lines drawn in an isometric drawing?
    a. 20°   c. 45°
    b. 30°   d. 60°

_____ 21. What type of pictorial drawing may have some accurately scaled dimensions?
    a. two-point perspective   c. oblique
    b. three-point perspective   d. all of the above

_____ 22. What type of pictorial will have most of its dimensions to an accurate scale?
    a. one-point perspective   c. three-point perspective
    b. two-point perspective   d. isometric

_____ 23. The horizon in a perspective drawing will always be level with the
    a. ground line   c. viewer's eye level
    b. bird's eye view   d. level of the picture plane

Chapter 20 Review Test

Directions: Write your answer to each question in the space provided.

_____ 1. Name at least three media used for renderings.
_____
_____
_____ 2. What is the term for a watercolor rendering when only black and gray tones are used?
_____ 3. Which medium is most effective for blending color gradations?
_____ 4. Which media are the most time consuming?
_____ 5. What is the term for preprinted screens used to add tones, textures, and shadows?
_____ 6. In adding shadows to an exterior perspective drawing, the position of which object must be established first?
_____ 7. What is the term for people or objects used in a drawing to enhance size, distance, or reality?
_____ 8. What two media can be effectively combined to provide fine detail and also add color to surrounding areas?
_____
_____ 9. In what way is the appearance of people in different locations changed to reflect where they are in relation to the picture plane?

Directions: In the space at the left, write the letter of the choice that best completes each statement.

____ 10. To render is to make a drawing
    a. more colorful   c. show perspective
    b. more realistic   d. with fewer features

____ 11. Felt markers
    a. blend easily   c. do not blend easily
    b. can be smudged   d. can be used for washes

____ 12. Ink lines can show texture when
   a. the distance between them is altered
   b. acrylics are added
   c. only smooth surfaces are shown
   d. all of the above

____ 13. A gradual change in shadow from light to dark usually indicates
   a. the sun is low in the sky
   b. ink was used
   c. a rounded object
   d. hidden features

____ 14. Location of shadows depends upon the direction and ___ of the light source.
   a. brightness
   b. duration
   c. angle
   d. size

____ 15. The smoothest surfaces are
   a. the most reflective
   b. the darkest
   c. the most shadowed
   d. difficult to show in perspective

____ 16. In preparing a rendering, which step is completed last?
   a. adding entourage
   b. adding textures
   c. emphasizing light and shadow
   d. darkening windows and overhangs

____ 17. Which medium can be used to add color to pencil renderings?
   a. ink
   b. pastels
   c. watercolors
   d. acrylics

____ 18. Which medium is used extensively for presentations and advertising?
   a. oil
   b. felt markers
   c. watercolors
   d. ink

____ 19. White watercolor or acrylic is often used in combination with
   a. colored illustration board
   b. oils
   c. overlays
   d. felt markers

____ 20. During daylight hours, windows usually appear
   a. in shadow
   b. open
   c. reflective
   d. dark

Chapter 21 Review Test

Directions: Place the steps in making a model in order by writing the correct letters in the spaces provided.

____ 1.   a. Apply siding materials or textures to walls.
____ 2.   b. Attach interior elevations to wall board.
____ 3.   c. Cut out all interior elevations.
____ 4.   d. Cut out doors and openings from interior partitions.
____ 5.   e. Attach exterior elevations to foam board.
____ 6.   f. Check plumb and squareness of walls.
____ 7.   g. Construct 3D built-ins and fixtures.
____ 8.   h. Determine base size from largest drawing.
____ 9.   i. Paint and install built-ins and fixtures.
____ 10. j. Attach floor plan to base.
____ 11. k. Cut roof panels from foam board.
____ 12. l. Cut exterior elevations from foam board.
____ 13. m. Construct removable roof assembly.
____ 14. n. Cut doors and windows from exterior elevations.
____ 15. o. Attach clear acetate for windows.
____ 16. p. Attach door and window trim to exterior walls.
____ 17. q. Cut out all doors from cardboard and attach with glue or tape.
____ 18. r. Glue interior partitions to base.
____ 19. s. Add outdoor areas (drives, walks, pool) to base.
____ 20. t. Glue exterior walls to base and to each other.
____ 21. u. Add landscape features.
____ 22. v. Add entourage.
____ 23. w. Paint interior walls and floors.

Directions: Write your answer to each question in the space provided.

_____   24. Name the two basic types of architectural models.

_____

_____   25. Name the three types of models used to check the form of a structure.

_____

_____

_____   26. Which type of model includes few details and is used only to check the layout and function of a design?

_____   27. Which type of model is composed of the framing members of a building?

_____   28. Which models usually include only the floor, three walls, and sometimes the ceiling of one room or area?

_____   29. Which type of model is used to promote the sale of land parcels?

_____   30. Which type of model is used to check construction methods?

_____   31. Which type of model shows the slope and shape of a site?

_____   32. Which type of presentation model is often a subject for photographs?

_____   33. Which type of model is used to check overall proportions?

Directions: Place the letter representing the correct answer on the left column line.

_____ 34. The layout model is
   a. used to check the overall size of the design
   b. usually made in solid form
   c. used to check the function of the design
   d. all of the above

_____ 35. The structural model is
   a. usually made in a solid form
   b. used to check the framing of the structure
   c. used to check the functional design of the building
   d. all of the above

_____ 36. Presentation models are used to
   a. promote sales
   b. show the shape and form of land parcels
   c. show community development projects
   d. all of the above

_____ 37. Detailed models are used to show the
   a. relationship with nearby buildings
   b. traffic patterns through a building
   c. structural aspects of a building
   d. final surface finishes of a building

_____ 38. Landform models are used to show the
   a. shape of a building site
   b. slopes of a building site
   c. contours of a building site
   d. all of the above

Chapter 22 Review Test

Directions: Write your answer to each question in the space provided.

_____ 1. Name the four basic structural materials used in today's buildings.

_____

_____

_____

_____ 2. Name the two basic structural types.

_____ 3. Name the four structural forces that exert stress on building materials.

_____

_____

_____

_____ 4. Which force tends to flatten objects?

_____ 5. Which force tends to pull objects apart?

_____ 6. Which force tends to make one part of an object slide past another?

_____ 7. Which force tends to twist an object out of shape?

_____ 8. What is the term for the weight of all movable objects on a building?

_____ 9. What is the term for the weight of building materials in a structure?

_____ 10. What term is used to describe forces on a building such as those from wind or an earthquake?

_____ 11. What is the term for bending caused by compression and tension?

_____ 12. Other than size, what can be changed about a material to increase its load-bearing capacity?

_____ 13. What is the term used for a horizontal member that is supported from only one side?

_____ 14. What is the term used to describe the distance between parallel structural members?

_____ 15. What is the term for the distance a horizontal member extends between vertical supports?

_____ 16. Snow on a roof is which type of load?

_____ 17. Roof shingles are which type of load?

_____ 18. What is the term for parts or sections constructed away from the building site?

_____ 19. What is the term for homes constructed at a factory and assembled on site?

Directions: Identify the types of forces acting on the structural members by writing the appropriate terms in the spaces provided.

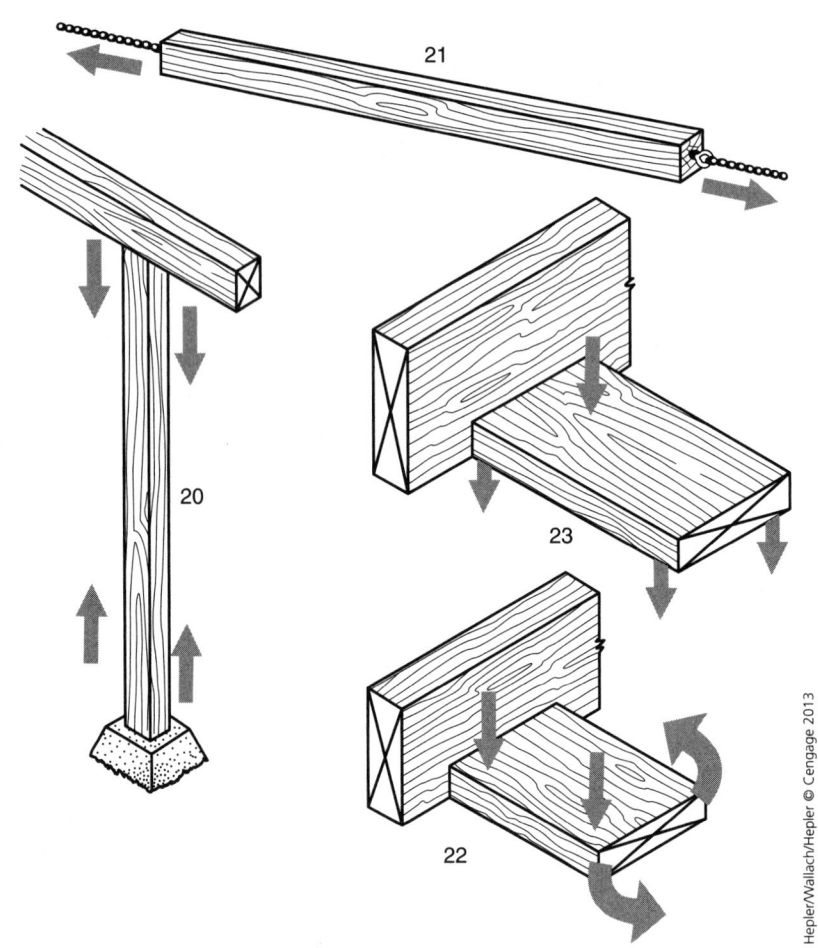

_____ 20.
_____ 21.
_____ 22.
_____ 23.

Chapter 25 Review Test

Directions: Write your answer to each question in the space provided.

_____ 1. Bricks are divided into which two categories?

_____

_____ 2. Name the three general types of concrete block.

_____

_____

_____ 3. What is the term used to describe the combination of sand, rocks, and other ingredients used to make concrete block?

_____ 4. Which type of structural tile can be exposed to weathering?

_____ 5. Name the four basic types of masonry wall construction.

_____

_____

_____

_____ 6. What is the term used to describe the patterns of arranging and attaching masonry in courses?

_____ 7. What unit is used to measure concrete's compressive strength?

_____ 8. What is the term used for compressing concrete so that its sides remain in compression during loading?

_____ 9. Name the method of counteracting stress by means of tendons stretched through cured concrete.

_____ 10. What is the term for a hollow steel column filled with concrete?

_____ 11. Which type of slab system is constructed by filling temporary forms with concrete on a temporary floor?

Directions: Write the names of the numbered items in the spaces provided.

_____
_____
_____
_____
_____
_____
_____
_____
_____
_____
_____
_____
_____
_____
_____
_____
_____
_____
_____
_____
_____
_____
_____

12.
13.
14.
15.
16.
17.
18.
19.
20.
21.
22.
23.
24.
25.
26.
27.
28.
29.
30.
31.
32.
33.
34.

Directions: Place the letter representing the correct answer on the left column line.

_____ 35. Which bond of masonry bricks has the most symmetrical pattern?
  a. basket-weave bond
  b. Ashlar bond
  c. stack bond
  d. running bond

_____ 36. Which bond of masonry bricks is asymmetrical?
  a. Ashlar bond
  b. running bond
  c. stack bond
  d. Flemish bond

_____ 37. A pretension concrete slab
  a. is reinforced with multiple rebars
  b. is reinforced with additional concrete
  c. is compressed with steel tendons
  d. will retain the concrete forms for additional strength

_____ 38. A typical concrete mix is
  a. 43% crushed rock, 27% sand, 3% water, 25% Portland cement, 2% air
  b. 41% crushed rock, 26% sand, 16% water, 11% Portland cement, 6% air
  c. 40% crushed rock, 30% sand, 10% water, 10% Portland cement, 10% air
  d. 40% crushed rock, 15% sand, 10% water, 40% Portland cement, 5% air

_____ 39. The spacing between a masonry-cavity wall is between
  a. 1″ and 1.5″
  b. 2″ and 3″
  c. 4″ and 5″
  d. 5″ and 6″

_____ 40. A lally column is a
  a. 4″ × 4″ wood column
  b. 6″ × 4″ wood column
  c. Masonry support column
  d. Steel pole column

_____ 41. Which laid brick pattern is in a vertical position?
  a. shiner
  b. sail
  c. soldier
  d. header

_____ 42. Which laid brick pattern is in a flat, horizontal, position?
  a. header
  b. soldier
  c. sailor
  d. shiner

_____ 43. How is the weight of bricks reduced?
   a. by adding air
   b. by adding insulation
   c. by adding holes
   d. by making them hollow
   e. by downsizing

_____ 44. Which of the following is (are) a masonry wall classification?
   a. solid
   b. cavity
   c. facing
   d. veneer
   e. all of the above

Chapter 26 Review Test

Directions: Write your answer to each question in the space provided.

_____ 1. Name the three general types of steel construction systems.
_____
_____

_____ 2. Which type of steel construction compares to wood skeleton-frame construction?

_____ 3. What is another term for steel girders?

_____ 4. Which vertical steel members rest on footings?

_____ 5. In which four shapes are steel bars available?
_____
_____
_____

_____ 6. Which structural steel member is rolled into a cross-section resembling the letter U?

_____ 7. Which structural steel shapes were formerly called I-beams?

_____ 8. Which structural steel beams are in the shape of an H?

_____ 9. Name the four basic methods for fastening steel members.
_____
_____
_____

_____ 10. Which type of architectural drawing shows the method and order of assembling each steel member?

Directions: Refer to the drawing on the next page. Write the names of the weld symbols in the spaces provided.

_____ 11.
_____ 12.
_____ 13.
_____ 14.
_____ 15.
_____ 16.
_____ 17.

_____ 18.
_____ 19.
_____ 20.
_____ 21.
_____ 22.
_____ 23.
_____ 24.
_____ 25.
_____ 26.
_____ 27.
_____ 28.
_____ 29.
_____ 30.
_____ 31.
_____ 32.
_____ 33.
_____ 34.
_____ 35.
_____ 36.
_____ 37.

_____ 38.
_____ 39.
_____ 40.
_____ 41.
_____ 42.
_____ 43.
_____ 44.
_____ 45.
_____ 46.
_____ 47.

Chapter 30 Review Test

Directions: Write the names of the pictured items in the spaces provided.

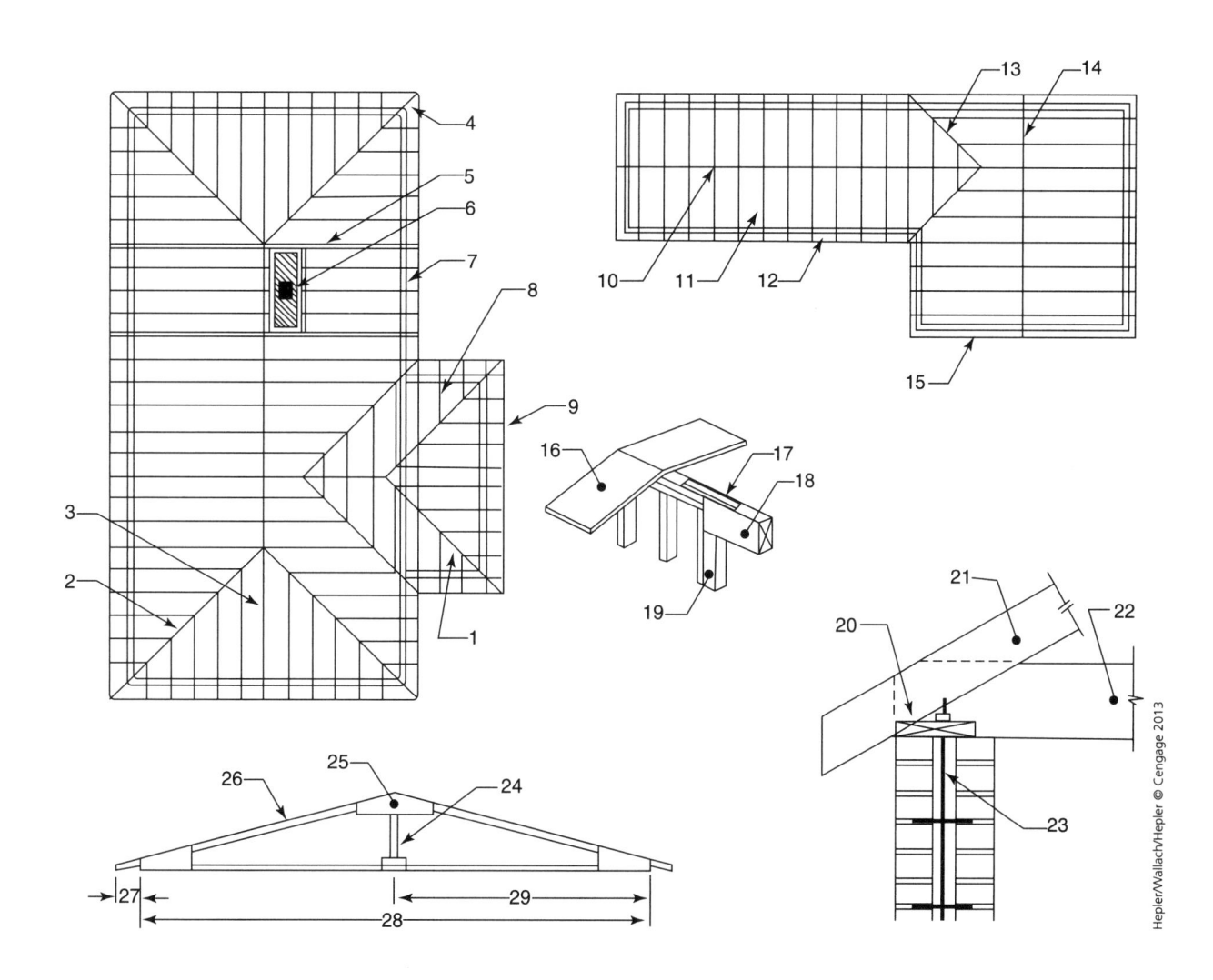

_____ 1.	_____ 9.
_____ 2.	_____ 10.
_____ 3.	_____ 11.
_____ 4.	_____ 12.
_____ 5.	_____ 13.
_____ 6.	_____ 14.
_____ 7.	_____ 15.
_____ 8.	_____ 16.

17.
18.
19.
20.
21.
22.
23.
24.
25.
26.
27.
28.
29.
30.
31.
32.
33.
34.
35.
36.
37.
38.
39.
40.
41.

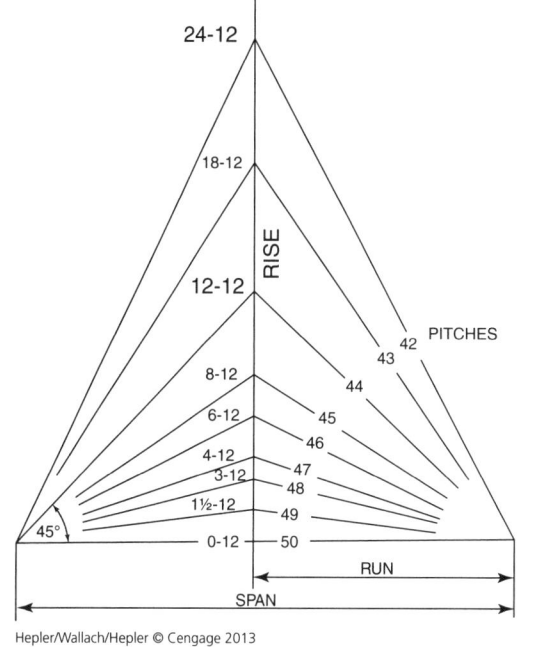

Directions: Place the letter representing the correct answer on the left column line.

_____ 42. What is the highest structural member in roof construction?
   a. cornice
   b. ridge board
   c. rafters
   d. soffit

_____ 43. What is the name of the detailed drawing where the top plate, rafters, and ceiling joists meet?
   a. cornice
   b. soffit
   c. purlin
   d. fascia
   e. ridge board

_____ 44. What structural member is *not* included in a roof truss?
   a. scab
   b. chord
   c. ridge board
   d. web
   e. gusset

_____ 45. What is a bird mouth cut?
   a. a notch in the top plate for a rafter
   b. a notch in the studs for a diagonal brace
   c. a bird feeder attached from a notch in the ridge board
   d. a notch in a rafter

_____ 46. What is a ledger strip?
   a. a vertical support member
   b. a horizontal support member
   c. a diagonal support brace
   d. all of the above

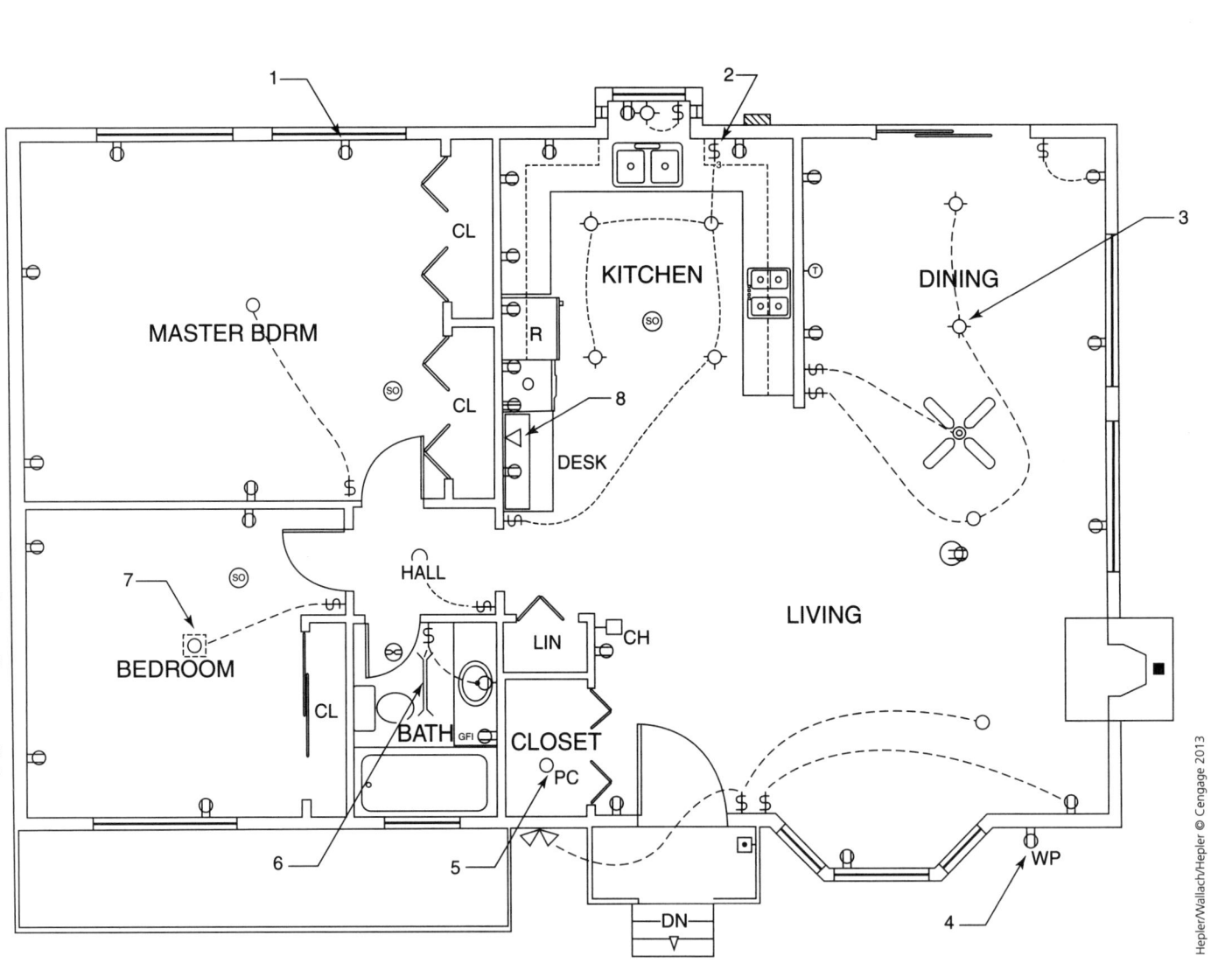

Chapter 31 Review Test

Directions: Write the names of the pictured items in the spaces provided.

_____ 1. _____ 5.

_____ 2. _____ 6.

_____ 3. _____ 7.

_____ 4. _____ 8.

Directions: Match the pictured items with the terms by writing the correct letters in the spaces provided. Note that some letters are capitals and others are not.

_____ 9. Single-pole switch
_____ 10. Double-pole switch
_____ 11. Three-way switch
_____ 12. Four-way switch
_____ 13. Weatherproof switch
_____ 14. Automatic door switch
_____ 15. Switch with pilot light
_____ 16. Low-voltage system switch
_____ 17. Circuit breaker
_____ 18. Ceiling outlet
_____ 19. Wall outlet
_____ 20. Ceiling outlet pull switch
_____ 21. Recessed light
_____ 22. Flood light
_____ 23. Spotlight
_____ 24. Vaporproof ceiling light
_____ 25. Fluorescent light
_____ 26. Telephone
_____ 27. Telephone jack
_____ 28. Buzzer
_____ 29. Chime
_____ 30. Push button

_____ 31. Bell
_____ 32. Double outlet
_____ 33. Single outlet
_____ 34. Triple outlet
_____ 35. Split-wire outlet
_____ 36. Weatherproof outlet
_____ 37. Floor outlet
_____ 38. Outlet with switch
_____ 39. Strip outlet
_____ 40. Heavy-duty outlet
_____ 41. Special-purpose outlet
_____ 42. Range outlet
_____ 43. Refrigerator outlet
_____ 44. Waterheater outlet

_____ 45. Garbage-disposal outlet
_____ 46. Dishwasher outlet
_____ 47. Iron outlet—pilot light
_____ 48. Washer outlet
_____ 49. Dryer outlet
_____ 50. Motor outlet
_____ 51. Electric door opener
_____ 52. Lighting distribution panel
_____ 53. Service panel
_____ 54. Junction box
_____ 55. Electric heater
_____ 56. Meter

Chapter 32 Review Test

Directions: Write your answer to each question in the space provided.

_____

1. Name the four main aspects of a home environment that NVAC systems control.

2. Which symbol is used to show the movement of air through ducts on an HVAC drawing?

3. Which symbol is used to show ducts that pass through a plane of projection?

4. By which method of transfer does heat flow as waves through space?

5. By which method of transfer does hot air rise and cooler air take its place?

6. By which method does heat move through a solid object?

7. What is the standard unit of measurement for heat *generated*?

8. What is the measure of heat *flow*?

9. What is the rating given to a building material that measures its resistance to heat flow?

10. Name the three types of forced-air furnaces.

11. What is another term for hot water heating systems?

12. Which HVAC device can both cool and heat buildings?

13. What is the term for a material that will absorb heat from the sun and later radiate that heat back into the air?

Directions: Match the terms to the pictured items by writing the correct letters in the spaces provided.

_____ 14. Blanket insulation

_____ 15. Compressor

_____ 16. Condenser

_____ 17. Direction of flow

_____ 18. Drain (floor)

_____ 19. Duct size

_____ 20. Duct-section flow

_____ 21. Duct-section return

_____ 22. Exhaust inlet

_____ 23. Forced convection

_____ 24. Fuel-oil return

_____ 25. Gas

_____ 26. General insulation

_____ 27. Hot-water heating return

_____ 28. Loose insulation

_____ 29. Pump

_____ 30. Pump—suction

_____ 31. Radiator—convectors

_____ 32. Reflective foil

_____ 33. Register

_____ 34. Steam boiler

_____ 35. Supply outlet

_____ 36. Switch

_____ 37. Thermometer

_____ 38. Thermostat

_____ 39. Valve, safety

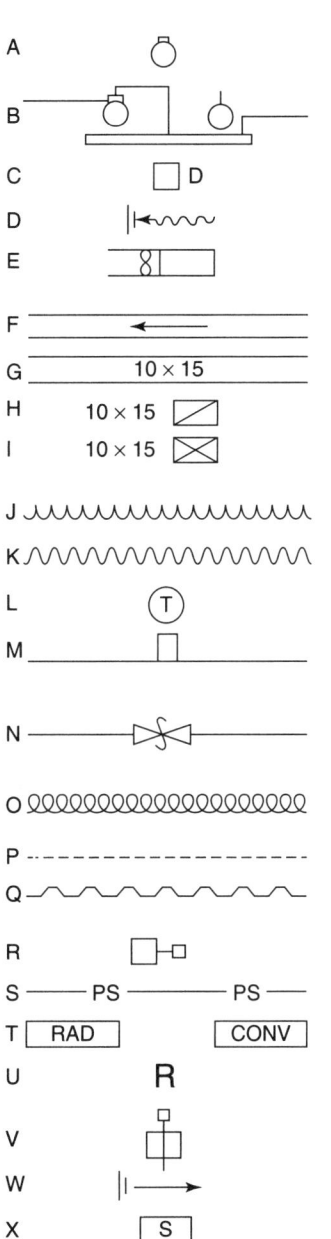

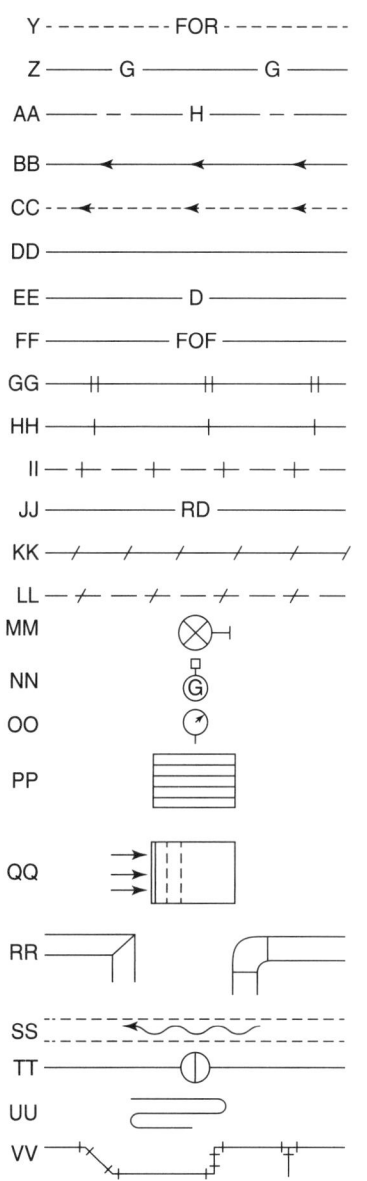

Chapter 33 Review Test

Directions: Write your answer to each question in the space provided.

_____

_____

_____

_____

_____

_____

_____

_____

1. What is another term for horizontal waste lines?
2. What is the term for the portion of the soil stack above the highest branch intersection?
3. What is the term for a large tank buried in the ground in which solid waste is processed by bacteria?
4. What is the term for the process of liquid absorption in a drainage field?
5. Name the five materials in which light construction plumbing pipes are available.

Directions: Write the names of the numbered items in the spaces provided.

_____ 6.

_____ 7.

_____ 8.

_____ 9.

_____ 10.

_____ 11.

_____ 12.

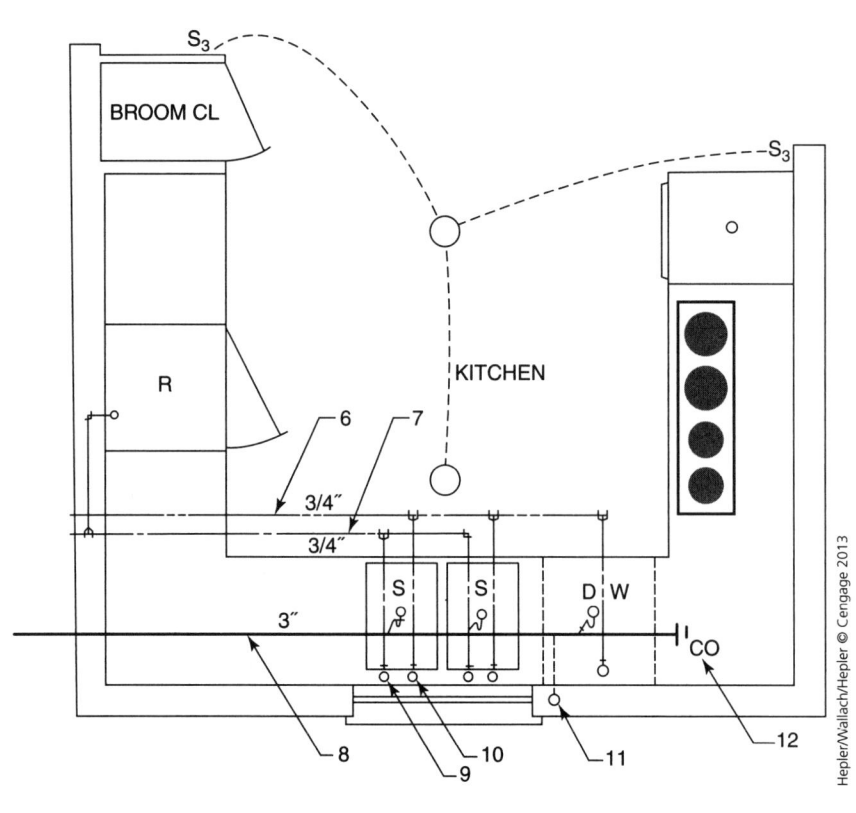

Chapter 33 Review Test ■ 259

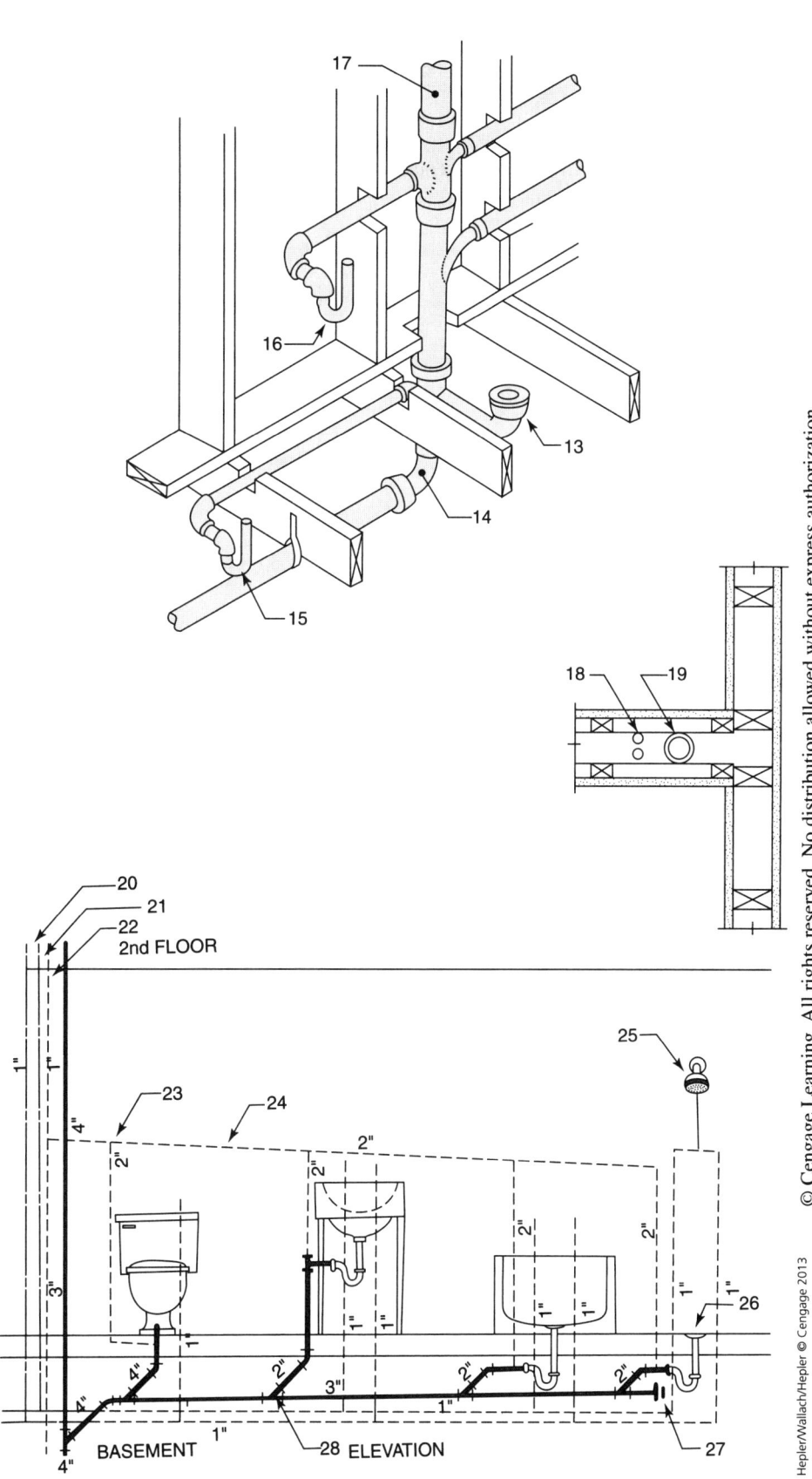

_____ 13.
_____ 14.
_____ 15.
_____ 16.
_____ 17.
_____ 18.
_____ 19.
_____ 20.
_____ 21.
_____ 22.
_____ 23.
_____ 24.
_____ 25.
_____ 26.
_____ 27.
_____ 28.

Chapter 34 Review Test

Directions: Write your answer to each question in the space provided.

_____ 1. In a large architectural firm, who checks drawings?

_____ 2. Which is the most difficult type of plan to read?

_____ 3. What must be included on a change order?

_____ 4. How is the original obsolete portion of a drawing identified?

_____ 5. What is the term for a transparent sheet on which details are photocopied and that is then attached to an original drawing?

Directions: In the space at the left, write the letter of the choice that best completes each statement.

_____ 6. Which of the following tools is especially helpful in keeping track of what has been checked on drawings?
  a. colored pencils
  b. gummed notes
  c. computer
  d. architect's scale

_____ 7. Which of the following is checked on a working drawing?
  a. code compliance
  b. materials
  c. projections of views
  d. all of the above

_____ 8. If drawings in a set contain conflicting information,
  a. building codes may be violated
  b. builders cannot determine which is correct
  c. designers may be fined by the contractor
  d. information on the last drawing completed is considered correct

_____ 9. Which is the best way to study a combination plan?
  a. Study only one element, such as the electrical part, at a time
  b. Ask the designer to separate it into components
  c. Study only one section of the structure at a time
  d. None of the above

_____ 10. Which of the following people *cannot* initiate a change order?
  a. client
  b. contractor
  c. building inspector
  d. subcontractor

_____ 11. Which symbol is used to identify a change on a drawing?
  a. circle
  b. triangle
  c. diamond
  d. square

_____ 12. Revisions are listed on a drawing sheet
  a. on the back
  b. after the building is completed
  c. from the top down
  d. from the bottom up

Chapter 35 Review Test

Directions: Write your answer to each question in the space provided.

_____ 1. What is the term for charts used to show detailed component information that cannot fit on a set of drawings?

_____ 2. Which document contains detailed information about building materials required for on-site construction?

_____ 3. Which document contains detailed instructions telling how materials are to be used?

_____ 4. Door and window key symbols are sometimes shown on many different drawings. On what drawing are they *always* shown?

_____ 5. Name the three main uses for materials lists.

_____

_____

_____ 6. If there is a difference between information on a drawing and information on a set of specifications, which information normally takes legal precedence?

_____ 7. What phrase is used to allow contractors to substitute equivalent materials and products?

_____ 8. Which verb is used in specifications to define an obligation?

_____ 9. What is the name of the recommended specification format that has 16 major divisions?

_____ 10. What can be used to show the actual material and color of items listed on furniture or covering schedules?

Directions: Place the letter representing the correct answer in the left column blank.

_____ 11. What is a schedule?
   a. a computer-generated computation
   b. a specification list
   c. a chart with information
   d. a listing of the structural engineering

_____ 12. What is needed to make a cost analysis for a building project?
   a. materials list
   b. CSI format
   c. purchasing forms
   d. detailed contract

_____ 13. What are the specifications (specs)?
   a. a cost analysis
   b. detailed written information
   c. Sweets catalog format
   d. schedules

Directions: Indicate on which documents the items can be found by writing the correct letters in the spaces provided. Some answers may be used more than once.

_____ 14. Heat sensors

_____ 15. Ceiling fan

_____ 16. Drawer slides

_____ 17. Number of studs

_____ 18. B

_____ 19. Deadbolt information

_____ 20. Beam data

_____ 21. Mullion sizes

_____ 22. Chaise

_____ 23. RH or LH

_____ 24. Cooktop

_____ 25. Roman faucet

_____ 26. Towel racks

_____ 27. Drapery fabric

_____ 28. Pedestal lavatory

_____ 29. Manufacturer's code

_____ 30. Wallpaper

a. Structural schedule

b. Window schedule

c. Door schedule

d. Materials list

e. Floor plans

f. Door hardware schedule

g. Cabinet hardware schedule

h. Plumbing hardware schedule

i. Hardware accessory schedule

j. Plumbing fixture schedule

k. Electrical fixture schedule

l. Security system schedule

m. Appliance schedule

n. Coverings schedule

o. Paint and finish schedule

p. Furniture and accessory schedule

Chapter 36 Review Test

Directions: Write your answer to each question in the space provided.

_____ 1. Name the four major factors that influence building costs.

_____ 2. Name the three methods of determining building costs.

_____ 3. What is the cost of a one-level, 30′ × 60′ residence at $80 per square foot?

_____ 4. What is the term for the initial money a home buyer pays to the seller?

_____ 5. What is the term for the legal contract for a housing loan?

_____ 6. What is another term for the amount of a loan minus interest?

_____ 7. What is the range of years over which a home loan is usually paid?

_____ 8. Which type of taxes applies to home ownership?

_____ 9. What is the term for the charges the buyer must pay the lender for the use of the money?

_____ 10. What is the term for costs incurred prior to actual construction?

_____ 11. What is the term for costs that must be paid before a buyer can take possession of a property?

_____ 12. What is the term for interest rates that remain the same over the life of a loan?

_____ 13. What is the term for interest rates that can vary over the life of a loan?

_____ 14. What is the acronym for principal, interest, taxes, and insurance on a property?

_____ 15. Loan principal, interest, taxes, and insurance should take no more than a certain portion of a buyer's income. What percentage do most institutional lenders require that a buyer not exceed?

_____ 16. What is the term for the action taken by a mortgage holder to reclaim a property for lack of payment?

_____ 17. What is the name of the account used when a lender collects funds from a buyer to pay taxes or other expenses?

_____ 18. What is the construction cost of a two-level house measuring 40′ × 60′ with 8′ ceilings, if building costs are $6 per cubic foot?

_____ 19. Which kind of labor represents the largest percentage of the total cost of a building?

Directions: Refer to the drawing and answer the questions in the spaces provided.

_____ 20. What is the area of this house?

_____ 21. At $90 per square foot, what is the cost of this house?

_____ 22. At $120 per square foot, what is the cost of this house?

_____ 23. At $150 per square foot, what is the cost of this house?

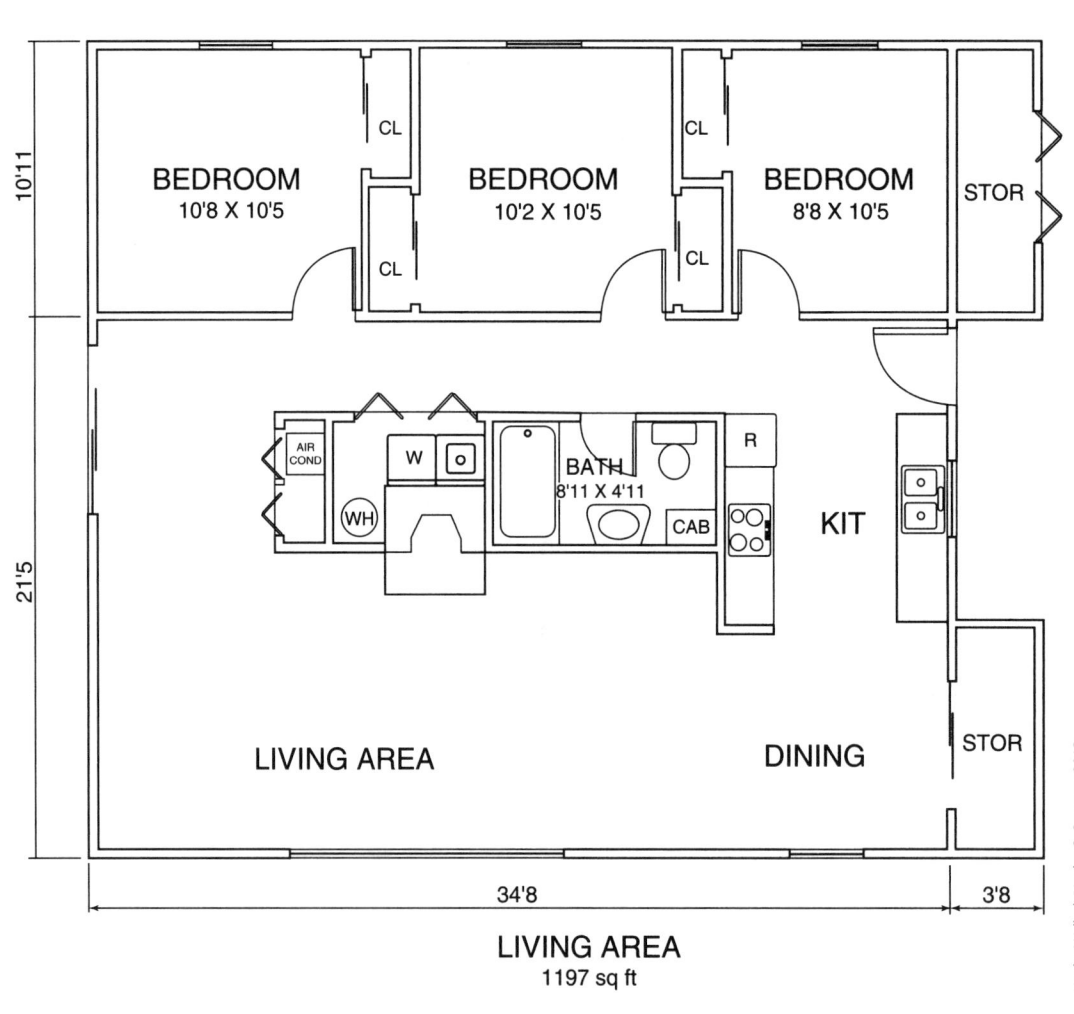

LIVING AREA
1197 sq ft

Chapter 37 Review Test

Directions: Write your answer to each question in the space provided.

_____ 1. What is the term for a collection of laws that ensure that building standards are met?

_____ 2. Which laws define and restrict the occupancy and use of buildings?

_____ 3. Which general codes cover load capacity of materials and structural integrity of construction?

_____ 4. Which codes cover only safety and performance requirements for a finished building?

_____ 5. Which codes include very specific requirements for the use and location of materials and methods of construction?

_____ 6. Which type of structural code would list permissible loads for a structure?

_____ 7. Which codes include such items as soil characteristics and water runoff?

_____ 8. Which code category specifies materials, sizes, and locations that are prohibited?

_____ 9. Which codes deal with disaster prevention?

_____ 10. Which document must be obtained from a local government *before* construction begins?

_____ 11. Which document is usually required *when* construction begins?

_____ 12. Which document is issued after a final inspection and before a building can be occupied?

_____ 13. What is the term for the legal agreement made between architect, builder, or owner?

_____ 14. Which document is used by contractors to guarantee that their work will be in accordance with the contract?

_____ 15. Which document guarantees that all material and service costs will be paid by the contractor?

_____ 16. What is the term for a legal certificate of property ownership?

_____ 17. Which legal document is used to take or hold the property of a debtor?

_____ 18. What is the term for a legal proposal from a contractor to construct a project as defined in a contract, in specifications, and in a set of drawings?

_____ 19. What is the term for a letter sent to contractors that outlines the requirements and conditions for submitting proposals to construct a project?

_____ 20. Most building codes are of which type?

Directions: Place the letter representing the correct answer in the left column blank.

_____ 21. Who is protected with the legal documents for a construction project?
   a. home owners
   b. contractors
   c. subcontractors
   d. architects
   e. general public
   f. all of the above

_____ 22. What does the local building code control?
   a. zoning
   b. structural elements
   c. health
   d. safety
   e. all of the above

_____ 23. What is *not* defined by zoning ordinances?
   a. size of structure
   b. location of structure
   c. health of occupants
   d. height of structure
   e. property line setbacks

_____ 24. What is *not* defined by site-related codes?
   a. soil percolation test
   b. soil support
   c. exterior surface finishes
   d. test boring in soil

_____ 25. What is *not* defined within the property deed?
   a. building codes
   b. zoning ordinances
   c. certificate of ownership
   d. easements
   e. cost analysis

APPENDIX REVIEW TEST

Directions: Name the type of worker responsible for the jobs listed below by writing your answers in the spaces provided.

_____ 1. Plans for the overall growth and development of cities.

_____ 2. Helps architects make models of structures.

_____ 3. Makes drawings having detailed dimensions for special parts of a design.

_____ 4. Prepares renderings.

_____ 5. Plans and designs all aspects of a building site.

_____ 6. Selects colors, fabrics, and window treatments for the inside of a structure.

_____ 7. Plans and designs structures under the direction of an architect.

_____ 8. Plans, designs, and directs large construction projects.

_____ 9. Specializes in designing the structural framework of a large building.

_____ 10. Designs and oversees installation of street lighting systems.

_____ 11. Uses engineer's notes to prepare drawings of a structure's framework.

_____ 12. Designs shapes, materials, and devices to suppress noise in a shopping mall.

_____ 13. Designs the operational parts of a large structure, such as a conveyor system.

_____ 14. Computes the cost of construction projects.

_____ 15. Analyzes an architect's plans and describes the exact type and quantity of materials to be used.

_____ 16. Describes the size, position, and topography of a piece of land.

_____ 17. Responsible for only part of a construction project, such as the plumbing system.

_____ 18. Shapes and smoothes fresh concrete surfaces.

_____ 19. Builds walls, chimneys, fireplaces, and arches.

_____ 20. Troubleshooter who keeps a building project on schedule.

Directions: Place the letter representing the correct answer in the left column

_____ 21. What profession does *not require* 4 or 5 years of college, an internship, and a state examination?
   a. architect
   b. landscape architect
   c. architectural designer
   d. civil engineer

_____ 22. What profession *does require* 4 or 5 years of college, an internship, and a state examination?
   a. structural engineer
   b. specification writer
   c. building inspector
   d. electrician specialist

_____ 23. What is the responsibility of an expediter?
   a. to keep building costs to a minimum
   b. to keep the building materials arriving on schedule
   c. to keep track of all the building inspections
   d. to ensure all construction follows the working drawings

_____ 24. Who is an HVAC technician?
   a. a technician who installs internal vacuum systems
   b. a heating and air conditioning technician
   c. a home valance and ceiling lighting specialist
   d. a hydronic vacuum appliance and circuit specialist